**Cover Crops and
Soil Ecosystem Services**

Cover Crops and Soil Ecosystem Services

Humberto Blanco
Department of Agronomy and Horticulture
University of Nebraska
Lincoln, NE, USA

Copyright © 2023 American Society of Agronomy, Inc. / Crop Science Society of America, Inc. / Soil Science Society of America, Inc. All rights reserved.

Copublication by American Society of Agronomy, Inc. / Crop Science Society of America, Inc. / Soil Science Society of America, Inc. and John Wiley & Sons, Inc.

No part of this publication may be reproduced, stored in a retrieval system, or transmitted in any form or by any means electronic, mechanical, photocopying, recording, scanning, or otherwise, except as permitted by law. Advice on how to reuse material from this title is available at http://wiley.com/go/permissions.

The right of Humberto Blanco to be identified as the author of this work has been asserted in accordance with law.

Limit of Liability/Disclaimer of Warranty
While the publisher and author have used their best efforts in preparing this book, they make no representations or warranties with respect to the accuracy of completeness of the contents of this book and specifically disclaim any implied warranties or merchantability of fitness for a particular purpose. No warranty may be created or extended by sales representatives or written sales materials. The publisher is not providing legal, medical, or other professional services. Any reference herein to any specific commercial products, procedures, or services by trade name, trademark, manufacturer, or otherwise does not constitute or imply endorsement, recommendation, or favored status by the ASA, CSSA and SSSA. The views and opinions of the author(s) expressed in this publication do not necessarily state or reflect those of ASA, CSSA and SSSA, and they shall not be used to advertise or endorse any product.

Editorial Correspondence:
American Society of Agronomy, Inc.
Crop Science Society of America, Inc.
Soil Science Society of America, Inc.
5585 Guilford Road, Madison, WI 53711-58011, USA

agronomy.org • crops.org • soils.org

Registered Offices:
John Wiley & Sons, Inc., 111 River Street, Hoboken, NJ 07030, USA

For details of our global editorial offices, customer services, and more information about Wiley products, visit us at www.wiley.com.

Wiley also publishes its books in a variety of electronic formats and by print-on-demand. Some content that appears in standard print versions of this book may not be available in other formats.

Library of Congress Cataloging-in-Publication Data
Names: Blanco, Humberto, 1961- author. | American Society of
  Agronomy, issuing body. | Crop Science Society of America, issuing body.
  | Soil Science Society of America, issuing body.
Title: Cover crops and soil ecosystem services / Humberto Blanco.
Description: First edition. | Hoboken, NJ, USA : Wiley-ACSESS, 2023. | Includes bibliographical
  references.
Identifiers: LCCN 2023002570 (print) | LCCN 2023002571 (ebook) | ISBN
  9780891186397 (hardback) | ISBN 9780891186427 (adobe pdf) | ISBN
  9780891186410 (epub)
Subjects: LCSH: Cover crops. | Soil protection. | Soil conservation.
Classification: LCC SB284 .B536 2023 (print) | LCC SB284 (ebook) | DDC
  631.4/52–dc23/eng/20230215
LC record available at https://lccn.loc.gov/2023002570
LC ebook record available at https://lccn.loc.gov/2023002571

Cover Design: Wiley
Cover Image: Courtesy of Humberto Blanco

Set in 9.5/12.5pt STIXTwoText by Straive, Pondicherry, India

Contents

Preface *xii*

1 Cover Crops and Soil Ecosystem Services *1*
1.1 Cover Crops *1*
1.2 Soil Ecosystem Services *4*
1.3 Cover Crops and Soil Ecosystem Services *6*
1.4 Summary *8*
     References *9*

2 Cover Crop Biomass Production *12*
2.1 Cover Crops and Biomass *12*
2.2 Aboveground Biomass Production *13*
2.2.1 Temperate Regions *13*
2.2.2 Semiarid Temperate Regions *14*
2.2.3 Tropical and Subtropical Regions *15*
2.3 Belowground Biomass Production *17*
2.4 Threshold Level of Biomass Production *18*
2.5 Management Practices that Affect Biomass Production *18*
2.5.1 Planting Early *19*
2.5.1.1 Interseeding *19*
2.5.1.2 Planting after Summer Crop Harvest *21*
2.5.1.3 Planting after Corn Silage or Short-Growing Season Crop Harvest *23*
2.5.2 Terminating Late *23*
2.5.3 Cover Crop Mixes *25*
2.5.4 Seeding Rate *26*
2.5.5 Planting Method *26*
2.5.6 Tillage and Cropping Systems *27*
2.5.7 Soil Texture and Number of Years in Cover Crops *28*

2.5.8	Irrigation and Fertilization	28
2.6	Summary	29
	References	30

3	**Soil Health**	**35**
3.1	Soil Health	35
3.2	Cover Crops and Soil Health	37
3.2.1	Soil Physical Properties	37
3.2.1.1	Soil Compaction	37
3.2.1.2	Soil Structure	39
3.2.1.3	Water Infiltration	40
3.2.1.4	Temperature	42
3.2.2	Soil Chemical Properties	42
3.2.3	Soil Biological Properties	43
3.2.3.1	Microorganisms	44
3.2.3.2	Macroorganisms	45
3.3	Interconnectedness of Soil Health Parameters	47
3.4	Managing Soil Health	47
3.4.1	Biomass Production	48
3.4.2	Time after Cover Crop Introduction	48
3.4.3	Cover Crop Species and Mixes	49
3.4.4	Tillage System	50
3.4.5	Initial Soil Condition	51
3.5	Summary	51
	References	52

4	**Water Erosion**	**58**
4.1	Overview	58
4.2	Runoff	59
4.3	Sediment Loss	62
4.4	Nutrient Loss	63
4.5	Soil Carbon Loss	63
4.6	A Leading Factor of Water Erosion: Biomass Production	64
4.7	Cover Crops and Erosion-Prone Systems	65
4.7.1	Low-Biomass Producing Cropping Systems	66
4.7.2	Corn Silage and Seed Corn	66
4.7.3	Crop Residue Removal	68
4.7.4	Orchard Crops	68
4.8	Summary	69
	References	69

5 Wind Erosion 73
5.1 Extent of Wind Erosion 73
5.2 Soil Loss 74
5.3 Soil Erodibility 76
5.4 Managing Wind Erosion 78
5.4.1 Biomass Production 78
5.4.2 Cover Crop Species 79
5.4.3 Growth Stage and Seeding Rate 80
5.4.4 Crop and Tillage Systems 81
5.4.5 Climate 82
5.5 Summary 82
References 83

6 Nutrient Losses 85
6.1 Implications of Nutrient Losses 85
6.2 Nutrient Leaching 86
6.3 Dissolved Nutrients in Runoff 89
6.4 Nutrient Release from Cover Crops 90
6.5 Management Implications 92
6.6 Nutrient Stratification 93
6.7 Summary 94
References 94

7 Soil Gas Emissions 97
7.1 Carbon and Nitrogen Emissions 97
7.2 Carbon Dioxide 98
7.3 Nitrous Oxide 99
7.4 Methane 100
7.5 Factors Affecting Soil Gas Emissions 100
7.5.1 Cover Crop Species 101
7.5.2 Biomass Production 102
7.5.3 Nitrogen Fertilization 103
7.5.4 Tillage and Cropping System 104
7.5.5 Measurement Time 105
7.5.6 Soil Texture and Climate 106
7.6 Summary 106
References 107

8 Carbon Sequestration 109
8.1 The Need for Carbon Sequestration 109
8.2 Rates of Carbon Sequestration 110

8.3	Topsoil Versus Subsoil Carbon Sequestration *112*
8.4	Managing Carbon Sequestration *114*
8.4.1	Biomass Production *114*
8.4.2	Cover Crop Species and Mixes *116*
8.4.3	Years after Cover Crop Adoption *117*
8.4.4	Initial Soil Carbon Level *118*
8.4.5	Tillage Systems *119*
8.4.6	Soil Texture *120*
8.4.7	Topographic Characteristics *120*
8.4.8	Climate *121*
8.5	Cropping System Carbon Footprint *122*
8.6	Strategies to Enhance Cover Crop Potential to Sequester Carbon *122*
8.7	Summary *124*
	References *125*
9	**Soil Water** *128*
9.1	Soil Water Management *128*
9.2	High Precipitation Regions *129*
9.3	Low Precipitation Regions *129*
9.4	Mechanisms of Soil Water Storage with Cover Crops *132*
9.5	Water Management *134*
9.5.1	Biomass Production *135*
9.5.2	Timing of Cover Crop Termination *136*
9.5.3	Tillage System *136*
9.5.4	Soil Texture *137*
9.6	Summary *138*
	References *138*
10	**Weed Management** *141*
10.1	Cover Crops and Weeds *141*
10.2	Weed Suppression *142*
10.3	Managing Weeds *144*
10.3.1	Biomass Production and Surface Cover *145*
10.3.2	Cover Crop Species and Mixes *146*
10.3.3	Tillage System *148*
10.3.4	Climate *148*
10.4	Summary *149*
	References *150*
11	**Soil Fertility** *151*
11.1	Soil Fertility Management *151*
11.2	Organic Matter *152*

11.2.1	Nitrogen 154
11.2.2	Nitrogen Scavenging 154
11.2.3	Reduction of Nitrogen Losses 154
11.2.4	Nitrogen Fixation 155
11.3	Phosphorus 156
11.4	Other Nutrients 158
11.5	Soil pH 158
11.6	Cation Exchange Capacity 160
11.7	Carbon to Nitrogen Ratio 161
11.8	Summary 163
	References 163
12	**Crop Yields** *167*
12.1	Multi-functionality of Cover Crops 167
12.2	Crop Yields 168
12.3	Climate 169
12.3.1	Cool and Warm Climates 169
12.3.2	Water-Limited Regions 169
12.4	Factors Affecting Crop Production 170
12.4.1	Cover Crop Species 170
12.4.2	Nitrogen Fertilization 171
12.4.3	Biomass Production 171
12.4.4	Planting Time and Method 172
12.4.5	Termination Timing 174
12.4.6	Cover Crop Mixes 174
12.4.7	Years after Cover Crop Adoption 175
12.4.8	Tillage Systems 176
12.4.9	Soil Texture 176
12.5	Summary 177
	References 177
13	**Grazing and Harvesting** *180*
13.1	Cover Crop Biomass Removal 180
13.2	Grazing 181
13.2.1	Soil Compaction 181
13.2.2	Water Infiltration 184
13.2.3	Soil Carbon Dynamics and Sequestration 185
13.2.4	Crop Yields 186
13.3	Minimizing Potential Grazing Impacts 187
13.3.1	Amount of Biomass Removal 188
13.3.2	Stocking Rate 189
13.3.3	Years under Grazing 189

13.3.4	Tillage	190
13.3.5	Soil Water Content	190
13.4	Harvesting	191
13.5	Grazing and Harvesting: An Added Benefit from Cover Crops?	193
13.6	Summary	194
	References	194

14	**Economics**	**197**
14.1	Cover Crops and Farm Profits	197
14.2	Economic Analysis	198
14.2.1	Grazing and Harvesting	199
14.2.2	Weed Suppression	201
14.2.3	Nitrogen Credit	202
14.2.4	Soil Carbon Credit	203
14.2.5	Crop Residue Harvesting	204
14.2.6	Valuation of Other Ecosystem Services	204
14.3	Site-Specificity of Economic Benefits	205
14.4	Summary	206
	References	206

15	**Adaptation to Extreme Weather**	**210**
15.1	Extreme Weather Events	210
15.2	Droughts	211
15.3	Floods	212
15.4	Precipitation Extremes	214
15.5	Dust Storms	217
15.6	Temperature Extremes	217
15.7	Soil Resilience	218
15.8	Summary	220
	References	221

16	**Opportunities, Challenges, and Future of Cover Crops**	**224**
16.1	Opportunities	224
16.1.1	Ecosystem Services	224
16.1.2	Biomass Production	227
16.1.3	Economics	228
16.1.4	Fluctuating Climates	229
16.2	Challenges	229
16.2.1	Ecosystem Services	230
16.2.1.1	Biomass Production	230
16.2.1.2	Soil Properties	231

16.2.1.3	Carbon Sequestration	*232*
16.2.1.4	Dissolved Nutrients	*232*
16.2.1.5	Crop Yields	*233*
16.2.1.6	Economics	*233*
16.2.1.7	Fluctuating Climates	*234*
16.3	Remaining Questions	*234*
16.4	The Future of Cover Crops	*236*
16.5	Summary	*237*
	References	*238*

Appendix I: Common and Scientific Names Used in the Book *241*

Preface

Literature is replete with peer-reviewed and non-peer-reviewed information on cover crops. Indeed, the number of publications on cover crops has exponentially increased in recent years. However, a book that integrates the potential soil ecosystem services provided by cover crops based on peer-reviewed research information is difficult to find in the midst of the abundant literature. Such integration can be helpful to better understand the ability of cover crops for maintaining or improving soil ecosystem services. Thus, the purpose of this book is to present readers with science-based information on cover crop implications on soil ecosystem services.

The positive or negative impacts of cover crops on soil ecosystem services are often overgeneralized in non-peer-reviewed literature and during general conversations. Also, cover crop experience from one location or region is often extrapolated to other agroecozones with different soil types, cover crop management scenarios, and climatic conditions without consideration of potential site-specific soil and crop response to cover crops. Discussion based on the research on both benefits and barriers surrounding cover crops can help with making informed decisions regarding cover crop adoption and management.

The discussion of cover crop impacts on soil ecosystem services in this book is based on published reviews, meta-analyses, and individual studies. When no reviews or meta-analyses are available for in-depth understanding of a given topic, the book cites relevant case studies or reports summaries of recent studies to support discussions and conclusions. Because most research work on cover crops has been conducted in temperate regions such as the U.S., most of the conclusions in this book are based on studies from these regions.

This book covers these topics: cover crop biomass production, soil properties, soil erosion, nutrient losses, soil gas emissions, soil C sequestration, soil water, weed management, nutrient management, crop production, and cover crop grazing and harvesting. It also includes some discussion on cover crop economics, cover crop potential for adaptation to extreme weather events, and opportunities and challenges of cover crop management. Factors that can affect cover crop impacts on soil ecosystem services are emphasized in each chapter. This book emphasizes 'soil' because understanding soils is critical to understand cover crops and their services.

This book is written for researchers, soil conservationists, crop consultants, students, and others interested in exploring the complex system changes that happen when cover crops are used in a cropping system. However, this book is not a guideline for cover crop establishment and management. The intention is to discuss cover crops and relevant soil ecosystem services based on experimental research information that advances our understanding of the potential of cover crops for addressing the declining soil ecosystem services and increasing concerns about weather extremes. It is my hope this book will further stimulate the discussion on the role of cover crops in the changing face of agriculture.

1

Cover Crops and Soil Ecosystem Services

1.1 Cover Crops

According to the Soil Science Society of America, cover crops are defined as a *"close-growing crop that provides soil protection, seeding protection, and soil improvement between periods of normal crop production, or between trees in orchards and vines in vineyards. When plowed under and incorporated into the soil, cover crops may be referred to as green manure crops"* (SSSA, 2022). Cover crops are not entirely new. Their use dates back over several millennia or probably to the origins of agriculture. Literature indicates that cover crops were used as green manure by civilizations in eastern Asia and ancient Rome approximately 3000 years ago (Groff, 2015; Lipman, 1912). Ancient civilizations used cover crops such as legumes as a source of essential nutrients to support soil fertility and productivity. In early times, cover crops were normally incorporated into soil to accelerate decomposition and improve soil fertility and thus were synonymous to green manure. In the U.S., Native Americans often used a mix of crops to improve crop diversity, which portrayed cover crop mixes (Groff, 2015). In the late 1700s, the first U.S. president, George Washington, was one of the first promoters of using cover crops to conserve soil in the Americas, and he often planted clover, grass, and buckwheat as cover crops (Groff, 2015). At the time, cover crops were mostly used in nutrient-depleted soils including monocultures of cotton in the southern U.S. and in other crops with limited residue input.

In the early 1900s, Hugh Hammond Bennett, known as the Father of Soil Conservation, vehemently advocated for the use of cover crops to reduce soil erosion, reduce nutrient leaching, and improve soil productivity during and in the aftermath of the Dust Bowl in his influential book "Soil Conservation" (Bennett, 1939). He considered cover crops as an integral piece to conserve soil and halt soil degradation. Indeed, Hugh Hammond Bennett was testifying before

---

Cover Crops and Soil Ecosystem Services, First Edition. Humberto Blanco.
© 2023 American Society of Agronomy, Inc. / Crop Science Society of America, Inc.
/ Soil Science Society of America, Inc. Published 2023 by John Wiley & Sons, Inc.

Congress in spring 1935 on the need to implement better soil conservation practices, such as cover crops, as a dust storm approached Washington, D.C. at the peak of the Dust Bowl. His testimony, coinciding with the dramatic arrival of the dust storm to the Capitol, facilitated the passage of the Soil Conservation Act by Congress and subsequent signing by President Roosevelt on April 27, 1935 (The National Agricultural Law Center, 1935). The Act specifically called for the implementation of soil erosion prevention measures such as growing vegetation (e.g., cover crops).

Cover crops were often used before World War II (1939–1945). Common cover crop species used by early adopters included crimson clover, field pea, crotalaria, sudangrass, millet, sweet clover, alfalfa, hairy vetch, winter rye, buckwheat, and others (Bennett, 1939; Lipman, 1912). Legume cover crops were used as sources of nutrients (e.g., N, C), while grass cover crops were used for erosion control. Following World War II, the rapid production and vast availability of commercial pesticides, herbicides, and synthetic fertilizers led to a slowed interest in the use of cover crops. Cover crop use was relatively minimal between 1940 and 1980 although organic farmers used cover crops throughout this period. However, increasing concerns over soil C losses, soil degradation, nutrient runoff, and nitrate leaching from agricultural lands contributed to reemergence of interest in cover crops in the 1980s.

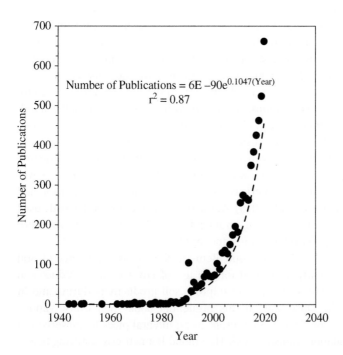

Figure 1.1 The number of publications on cover crops has increased exponentially in the past few decades. Web of Science.

The heightened interest in using cover crops in the past few decades resulted in an exponential increase in the number of publications (Figure 1.1). A search in Web of Science using the phrase "cover crops" up to December 2021, shows the number of publications was only 35 prior to 1980, 613 between 1980 and 2000, and 5967 between 2001 and 2021 (Figure 1.1). Most articles published before 1980 discussed the use of cover crops in low-biomass producing crops (e.g., cotton), orchards (e.g., cover crops planted under or between trees), and vegetable gardens for pest suppression (e.g., nematodes). Between 1980 and 2000, the main reasons for the use of cover crops were water and wind erosion control, soil fertility improvement, and the suppression of pests and diseases, while between 2000 and 2010, there was greater discussion of cover crops for sequestering soil C and improving soil quality.

Research on cover crops during the last decade (2010–2020) has expanded beyond the on-farm benefits from cover crop use. Now, most publications focus on ecosystem services or multi-functionality of cover crops, climate mitigation potential, agricultural intensification, soil biological environment, soil water management, and the challenges and opportunities of cover crop management. Also, several publications have recently emerged regarding the potential of cover crops to support livestock production via grazing or haying while improving farm economics and maintaining soil ecosystem services.

The above chronology indicates that while the use of cover crops is nothing new, interest in the multi-functionality of cover crops has increased in recent years (Figure 1.2).

Figure 1.2   No-till sunn hemp (left) and late-maturing soybean (right) summer cover crops in winter wheat-grain sorghum systems for enhancing soil ecosystem services. Blanco-Canqui et al., 2011; Photo by H. Blanco.

Most research on cover crops has been conducted in the U.S. Many studies are also evaluating management strategies to make cover crops work in water-limited environments where cover crop success can be restricted due to limited precipitation. Despite abundant literature, the adoption rate of cover crops is still slow. For example, in the U.S., cover crops are used in less than 5% of croplands although the adoption rate depends on the region. In some regions, cover crops are used in approximately 20% of croplands (Yoder et al., 2021).

1.2 Soil Ecosystem Services

Soils provide many invaluable services to humans (Figure 1.3). Not only do soils support food crops, but soils also support biomass as fiber for the textile industry,

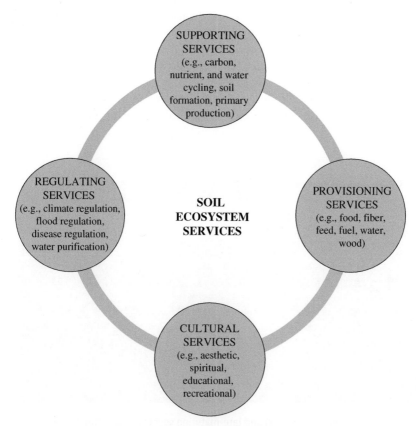

Figure 1.3 Soils provide numerous essential services. MEA, 2005; Dominati et al., 2010; Comerford et al., 2013.

feedstock for biofuel production, and forage for animals (Hatfield et al., 2017). Soils capture precipitation and irrigation water, clean water, degrade pollutants, sequester atmospheric C, adsorb and retain nutrients, moderate temperature, provide habitat for billions of soil organisms, and deliver many other services. These essential soil services can be grouped into four categories: supporting, provisioning, regulating, and cultural services (Figure 1.3; MEA, 2005; Dominati et al., 2010; Comerford et al., 2013). Supporting services refer to C, nutrient, and water cycling as well as primary production, soil formation, and microbial habitat, while provisioning services refer to the products we obtain from soil including water, food, fiber, feed, and fuel (MEA, 2005). Soils do not simply support and deliver products but also mediate and regulate many processes, which are vital to plants, animals, and humans (Hatfield et al., 2017). Such services are considered as regulating services and include climate regulation, air quality regulation, water movement and purification, prevention of floods, and management of pests and diseases (MEA, 2005). Also, soils have aesthetic, spiritual, educational, and recreational value, and these are grouped as cultural services (MEA, 2005).

The supporting, provisioning, regulating, and cultural services are all interconnected and subject to feedbacks among services. As an example, a soil may not be able to produce food and biomass if it cannot effectively cycle and recycle water and nutrients. Perhaps the leading service from the soil is the supporting service, which directly affects the capacity of the soil to produce food, fiber, feed, and fuel (provisioning services), buffer or moderate temperature, and contribute to water flow and storage (regulating services), improve landscape esthetics, and serve as recreational, educational, and spiritual space (cultural services).

The concept of soil ecosystem services is often implied but not entirely valued (Pires-Marques et al., 2021; Yee et al., 2021). Any service in society has a value. Thus, the services that soils provide have a value. Assigning a monetary value on different soil ecosystem services has been the topic of recent publications (Comerford et al., 2013; Mikhailova et al., 2021; Pires-Marques et al., 2021). Provisioning services such as food and biomass production can be easily valued because these products are marketable, but how about the rest of soil ecosystem services? For example, what is the monetary value of clean water, clean air, reduced C losses, reduced sediment losses, and other services?

Some have quantified the value of select soil ecosystem services by considering the "avoided cost" of soil erosion (Pires-Marques et al., 2021) and CO_2 emissions (Mikhailova et al., 2021), while others considered natural capita and flow of ecosystem services for the economic valuation of such services (Comerford et al., 2013; Yee et al., 2021). Comerford et al. (2013) reported some estimates of economic values for different ecosystem services including nitrate leaching, sediment loss, nutrient cycling, soil formation, salinization, contamination, and others. Most of these estimates are for supporting services. Also, available approaches often value a single service such as C sequestration or reduced losses of C as CO_2 (Mikhailova et al., 2021). However, linkages of a given service with related ecosystem service indicators need further consideration (Comerford et al., 2013). For instance, if a soil sequesters C or reduces CO_2 emissions, then soil aggregation or the amount of stable soil aggregates that contribute to the protection of C within aggregates should be considered during the valuation of services. This and other similar inter-related processes complicate the valuation.

Qualitative valuation of ecosystem services is relatively simple but quantitative valuation (monetary value) of ecosystem services is complex, especially when processes are interconnected or not directly marketable. A more comprehensive economic assessment of all indirectly marketable soil ecosystem services can help with decision-making process for the management of natural resources and thus ecosystem services. Indirectly marketable soil ecosystem services including recreation, spiritual fulfillment, landscape esthetics (e.g., year-round growing vegetation), mental and overall plant, animal, human health are often subtle but these soil services can be as valuable as marketable soil ecosystem services (Comerford et al., 2013; Yee et al., 2021).

The consideration of benefits from soil within the framework of ecosystem services is a holistic approach to view the soil as a service provider and one that deserves attention and care. Process-based models or quantitative frameworks are being developed to account for multiple soil processes contributing to a given service, although more refinement of such models is needed to fully quantify and value soil ecosystem services at farm-scales (Yee et al., 2021). It is clear that an understanding of the value of soil will be incomplete until we fully assign a quantitative value to each ecosystem service that soils provide. However, a value cannot be assigned until the impacts of cover crops on each soil service is quantified and understood.

1.3 Cover Crops and Soil Ecosystem Services

The concept of soil ecosystem services emerged in recent decades due to declining services from the soil and the need to improve, maintain, and restore such services (Figure 1.3). The fate and downfall of many past civilizations depended on the ability of soils to continuously deliver vital services to plants, animals, and humans (Bennett, 1939). Increased erosion (Thaler et al., 2021), increased water pollution (Haque, 2021), development of hypoxic zones (Anderson et al., 2021), and other environmental problems are current signs of accelerated loss of ecosystem services from soils (Table 1.1). This is particularly true under increasing extreme weather events with intense rainstorms, frequent droughts, extreme temperatures, and heat waves. Thus, the challenge of this century is to ensure that soil ecosystem

Table 1.1 Some of the Current Signs Showing Soil Ecosystem Services Have Declined

Decline in soil ecosystem services	Source
Increased losses of soil C via erosion, leaching, and as C emissions	Minasny et al., 2017; Jian et al., 2020
Reduced water quality or increased water pollution	Blanco-Canqui, 2018; Haque, 2021
Increased hypoxic and anoxic events in lakes and coastal areas	Fennel and Testa, 2019; Anderson et al., 2021
Increased susceptibility to water erosion (increased runoff and sediment loss)	Fenta et al., 2020; Thaler et al., 2021
Increased minimum and maximum temperature or temperature extremes	Kaye and Quemada, 2017; Zscheischler and Fischer, 2020
Increased susceptibility to prolonged and frequent flooding	Kaye and Quemada, 2017; Wright et al., 2017
Reduced soil and agroecosystem resilience against droughts	Vogel et al., 2019; Zscheischler and Fischer, 2020
Overall reduced health of soils against extreme events	Lehmann et al., 2020

services are maintained or improved not only to meet the demands for food, fuel, fiber, and feed but also to reduce water pollution, air pollution, soil C loss, soil erosion, and others. Soil ecosystem services are finite and exhaustible as the soil is highly dynamic and susceptible to rapid degradation when not managed properly. Management determines the ability or inability of the soils to provide the essential provisioning, regulating, supporting, and cultural ecosystem services.

One of the reemerging biological strategies that has potential to improve and maintain soil ecosystem services from agricultural lands is the inclusion of cover crops into current cropping systems (Figure 1.2). Unlike other management practices such as the introduction of perennial vegetation (e.g., grass hedges) to croplands, cover crops would not compete with land for food production as they are often grown during times when no crops are growing in the field. Even when cover crops are interseeded along main crops or before the main crop harvest, cover crops do not appear to compete with the main crops nor reduce crop yields under proper management. Interseeding cover crops via aerial broadcasting or drilling with improved high clearance equipment when main crops are in the field is a subject of current research (Blanco-Canqui et al., 2017).

The question is: Can cover crops under different scenarios of cover crop management improve or enhance all the ecosystem services that soils provide? If not, how can the potential of cover crops to deliver soil ecosystem services be enhanced? It is often considered that cover crops would improve soil properties, sequester C, and improve other ecosystem services. In some cases, this common belief may, however, contrast with field research data. Adoption and management of cover crops may not be free of challenges (Roesch-McNally et al., 2018). A need exists to better understand the extent to which cover crops can maintain or enhance the multiple ecosystem services of agricultural lands based on experimental data.

Furthermore, many recent publications are emphasizing the multi-functionality of cover crops (Schipanski et al., 2014; Blanco-Canqui et al., 2015; Finney & Kaye, 2017). For instance, grazing or harvesting cover crops is generating interest (Franzluebbers & Stuedemann, 2008; Kelly et al., 2021). However, can cover crops be grazed or harvested and still be considered cover crops? The existing definition of cover crops does not appear to account for some of the potential multi-functionality of cover crops such as supporting livestock production (SSSA, 2022).

This book discusses how cover crops affect the numerous ecosystem services that soils provide under different cover crop management scenarios and climatic conditions based on experimental data. It also highlights challenges and opportunities with cover crops to manage soil ecosystem services. The ecosystem services are discussed in terms of soil health, water erosion, wind erosion, greenhouse gas emissions, C sequestration, nutrient losses, soil water, weed management, soil fertility, crop yields, and economics, among others. It also includes discussion on how grazing or harvesting of cover crops could alter the main purpose of cover crops, which is soil conservation and management.

1.4   Summary

Interest in growing cover crops is reemerging as one of the options to address the decline in soil ecosystem services from agricultural lands. Soil ecosystem services refer to the numerous benefits we receive from soils. Soils not only produce food and biomass (marketable services) but also filter water, sequester C, recycle nutrients, suppress weeds and diseases, and moderate soil temperature, among other services. These services are grouped into four categories: supporting, provisioning, regulating, and cultural services.

Cover crops can be a strategy to restore, improve, and maintain these essential services from soil. The use of cover crops dates back over three millennia but slowed in the mid-1900s due to the advent of inorganic fertilizers, herbicides, and pesticides after World War II. Interest rapidly increased after the 1980s due to heightened concerns of water pollution (e.g., hypoxia, anoxia), soil erosion, soil C losses, frequency of extreme weather events, and others. In early years, cover crops were primarily used for pest suppression and soil fertility improvement as green manure. Now, cover crops are being considered more and more as multi-functional systems that can deliver multiple soil ecosystem services. However, the potential of cover crops to function as multi-functional systems is not yet well understood. The question is: Can cover crop improve all soil ecosystem services? This book addresses this question based on experimental data on the impacts of cover crop management on soil environment, water quality, greenhouse gas emissions, C sequestration, soil water, weeds, crop yields, livestock production (e.g., grazing, haying), and other soil services.

References

Anderson, H. S., Johengen, T. H., Miller, R., & Godwin, C. M. (2021). Accelerated sediment phosphorus release in Lake Erie's central basin during seasonal anoxia. *Limnology and Oceanography, 66*, 3582–3595.

Bennett, H. H. (1939). *Soil Conservation* (993 p). McGraw-Hill Book Co., Inc.

Blanco-Canqui, H. (2018). Cover crops and water quality. *Agronomy Journal, 110*, 1633–1647.

Blanco-Canqui, H., Mikha, M., Presley, D. R., & Claassen, M. M. (2011). Addition of cover crops enhances no-till potential for improving soil physical properties. *Soil Science Society of America Journal, 75*, 1471–1482.

Blanco-Canqui, H., Shaver, T. M., Lindquist, J. L., Shapiro, C. A., Elmore, R. W., Francis, C. A., & Hergert, G. W. (2015). Cover crops and ecosystem services: Insights from studies in temperate soils. *Agronomy Journal, 107*, 2449–2474.

Blanco-Canqui, H., Sindelar, M., & Wortmann, C. S. (2017). Aerial interseeded cover crop and corn residue harvest: Soil and crop impacts. *Agronomy Journal, 109*, 1344–1351.

Comerford, N. B., Franzluebbers, A. J., Stromberger, M. E., Morris, L., Markewitz, D., & Moore, R. (2013). Assessment and evaluation of soil ecosystem services. *Soil Horizons, 54*, 1–14.

Dominati, E., Patterson, M., & MacKay, A. (2010). A framework for classifying and qualifying the natural capital and ecosystem services of soils. *Ecological Economics, 69*, 1858–1868.

Fennel, K., & Testa, J. M. (2019). Biogeochemical controls on coastal hypoxia. *Annual Review of Marine Science, 11*, 105–130.

Fenta, A. A., Tsunekawa, A., Haregeweyn, N., Poesen, J., Tsubo, M., Borrelli, P., Panagos, P., Vanmaercke, M., et al. (2020). Land susceptibility to water and wind erosion risks in the East Africa region. *Science of the Total Environment, 703*, 135016.

Finney, D. M., & Kaye, J. P. (2017). Functional diversity in cover crop polycultures increases multifunctionality of an agricultural system. *Journal of Applied Ecology*, *54*, 509–517.

Franzluebbers, A. J., & Stuedemann, J. A. (2008). Soil physical responses to cattle grazing cover crops under conventional and no tillage in the southern Piedmont USA. *Soil and Tillage Research*, *100*, 141–153.

Glossary of Soil Science Terms. Soil Science Society of America (SSSA). (2022). https://www.soils.org/publications/soils-glossary.

Groff, S. (2015). The past, present, and future of the cover crop industry. *Journal of Soil and Water Conservation*, *70*, 130A–133A.

Haque, S. E. (2021). How effective are existing phosphorus management strategies in mitigating surface water quality problems in the U.S.? *Sustainability*, *13*, 6565.

Hatfield, J. L., Sauer, T. J., & Cruse, R. M. (2017). Soil: The forgotten piece of the water, food, energy nexus. *Advances in Agronomy*, *143*, 1–46.

Jian, J., Du, X., Reiter, M. S., & Stewart, R. D. (2020). A meta-analysis of global cropland soil carbon changes due to cover cropping. *Soil Biology and Biochemistry*, *143*, 107735.

Kaye, J. & Quemada, M. (2017). Using cover crops to mitigate and adapt to climate change. A review. *Agronomy for Sustainable Development, 37*, 4.

Kelly, C., Schipanski, M. E., Tucker, A., Trujillo, W., Holman, J. D., Obour, A. K., Johnson, S. K., Brummer, J. E., Haag, L., & Fonte, S. J. (2021). Dryland cover crop soil health benefits are maintained with grazing in the U.S. High and Central Plains. *Agriculture, Ecosystems and Environment*, *313*, 107358.

Lehmann, J., Bossio, D. A., Kögel-Knabner, I., & Rillig, M. C. (2020). The concept and future prospects of soil health. *Nature Reviews Earth and Environment*. http://dx.doi.org/10.1038/s43017-020-0080-8

Lipman, J. G. (1912). *The Associative Growth of Legumes and Nonlegumes*. Bulletin No. 253. New Jersey Agricultural Experiment Station.

MEA (Millennium Ecosystem Assessment). (2005). *Ecosystems and Human Well-Being: Scenarios*. World Resources Institute.

Mikhailova, E. A., Zurqani, H. A., Post, C. J., Schlautman, M. A., Post, G. C., Lin, L., & Hao, Z. (2021). Soil carbon regulating ecosystem services in the state of South Carolina, USA. *Land*, *10*, 309.

Minasny, B., Malone, B. P., McBratney, A. B., Angers, D. A., Arrouays, D., Chambers, A., Chaplot, V., Chen, Z. S., Cheng, K., Das, B. S., Field, D. J., Gimona, A., Hedley, C. B., Hong, S. Y., Mandal, B., Marchant, B. P., Martin, M., McConkey, B. G., Mulder, V. L., ... Winowiecki, L. (2017). Soil carbon 4 per mille. *Geoderma*, *292*, 59–86.

Pires-Marques, É., Chaves, C., & Pinto, L. M. C. (2021). Biophysical and monetary quantification of ecosystem services in a mountain region: The case of avoided soil erosion. *Environment, Development and Sustainability*, *23*, 11382–11405.

Roesch-McNally, G., Basche, A., Arbuckle, J., Tyndall, J., Miguez, F., Bowman, T., & Clay, R. (2018). The trouble with cover crops: Farmers' experiences with overcoming barriers to adoption. *Renewable Agriculture and Food Systems, 33*, 322–333.

Schipanski, M. E., Barbercheck, M., Douglas, M. R., Finney, D. M., Haider, K., Kaye, J. P., Kemanian, A. R., Mortensen, D. A., Ryan, M. R., Tooker, J., & White, C. (2014). A framework for evaluating ecosystem services provided by cover crops in agroecosystems. *Agricultural Systems, 125*, 12–22.

Thaler, E. A., Larsen, J., & J., and Q. Yiu. (2021). The extent of soil loss across the US Corn Belt. *Proceedings of the National Academy of Sciences of the United States of America, 118*, e1922375118.

The National Agricultural Law Center. (1935). Soil Conservation Act. http://nationalaglawcenter.org/wp-content/uploads/assets/farmbills/soilconserv1935.pdf

Vogel, E., Donat, M. G., Alexander, L. V., Meinshausen, M., Ray, D. K., Karoly, D., & Frieler, K. (2019). The effects of climate extremes on global agricultural yields. *Environmental Research Letters, 14*, 054010.

Wright, A. J., de Kroon, H., Visser, E. J. W., Buchmann, T., Ebeling, A., Eisenhauer, N., Fischer, C., Hildebrandt, A., Ravenek, J., Roscher, C., Weigelt, A., Weisser, W., Voesenek, L. A. C. J., & Mommer, L. (2017). Plants are less negatively affected by flooding when growing in species-rich plant communities. *New Phytologist, 213*, 645–656.

Yee, S. H., Paulukonis, E., Simmons, C., Russell, M., Fulford, R., Harwell, L., & Smith, L. M. (2021). Projecting effects of land use change on human well-being through changes in ecosystem services. *Ecological Modelling, 440*, 109358.

Yoder, L., Houser, M., Bruce, A., Sullivan, A., & Farmer, J. (2021). Are climate risks encouraging cover crop adoption among farmers in the southern Wabash River basin? *Land Use Policy, 102*, 105268.

Zscheischler, J., & Fischer, E. M. (2020). The record-breaking compound hot and dry 2018 growing season in Germany. *Weather and Climate Extremes, 29*, 100270.

2

Cover Crop Biomass Production

2.1 Cover Crops and Biomass

Can cover crops deliver numerous soil ecosystem services? One of the key drivers that determines the delivery of such services is the amount of cover crop biomass produced (Figure 2.1). Cover crop benefits may not be observed unless cover crop biomass production is high enough or reaches a threshold level that alters a service. Also, expanded uses of cover crops such as grazing and haying or harvesting for livestock or biofuel production depend on the amount of biomass produced to determine whether cover crops can be safely grazed or harvested without compromising the benefits of cover crops for other services.

A knowledge of both aboveground and belowground (root) biomass production is important to assess any changes in soil ecosystem services from cover crops. Indeed, some services such as soil C sequestration could depend more on belowground biomass production than on aboveground biomass production (Xu et al., 2021). Often, we focus on what we see in the field when cover crops are growing, which is aboveground biomass, and not on the "hidden" portion, which is belowground biomass. Also, interest in cover crop diversity or mixes is growing, but do multi-species cover crop mixes produce more aboveground and belowground biomass than monocultures? Some hypothesize that diverse cover crop mixes could concomitantly translate into delivery of diverse and multiple ecosystems services.

This chapter discusses the amount of aboveground and belowground biomass that cover crops can produce. It also discusses how different cover crop management strategies including planting and termination timing, seeding rates, mixes and monocultures, and other factors affect cover crop biomass production.

Cover Crops and Soil Ecosystem Services, First Edition. Humberto Blanco.
© 2023 American Society of Agronomy, Inc. / Crop Science Society of America, Inc. / Soil Science Society of America, Inc. Published 2023 by John Wiley & Sons, Inc.

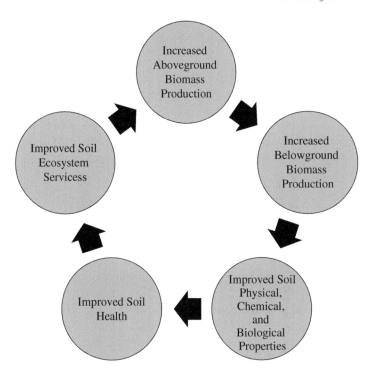

Figure 2.1 Interconnectedness among cover crop biomass production, soil properties, and soil ecosystems services.

2.2 Aboveground Biomass Production

The amount of cover crop biomass produced is highly variable as there are numerous factors that affect cover crop biomass production. Perhaps the most critical factors for cover crop production are precipitation and temperature. Cover crop biomass production varies not only from location to location but also from year to year within the same location due to differences in precipitation amount and other factors. Reduced growing degree days, limited soil moisture for cover crop germination, delayed germination, and other challenges can result in reduced cover crop biomass production. In water-limited regions, cover crop establishment is a major hindrance for biomass production.

2.2.1 Temperate Regions

Cover crops can produce, on average, 3.78 ± 3.08 Mg ha^{-1} of biomass in temperate regions (annual precipitation >750 mm; Ruis et al., 2019). Cover crops can produce lower amounts of biomass in colder temperate regions than in warmer

Table 2.1 Case Studies of Low Cover Crop Biomass Production When Cover Crops Are Seeded Late and Terminated Early Under Typical Corn and Soybean Systems in Cool Temperate Regions

Location	Crop	Cover crop	Planting time	Termination time	Cover crop biomass (Mg ha^{-1})
[a]Iowa, U.S.	Soybean	Winter rye	Mid-fall	Early spring	0.6
[b]Nebraska, U.S.	Continuous corn	Winter rye	Mid-fall	Early spring	0.8
[c]Nebraska, U.S.	Corn–soybean	Winter rye	Mid-fall	Early spring	1.1
[d]Missouri, U.S.	Continuous corn	Winter rye	Mid-fall	Mid-spring	0.6

[a] Moore et al. (2014).
[b] Sindelar et al. (2019).
[c] Koehler-Cole et al. (2020).
[d] Rankoth et al. (2019).

temperate regions due to shorter growing cover crop seasons in colder regions. Within temperate regions, cover crop biomass production follows this order of warm > mild > cold. In cool temperate regions, cover crop biomass production may not exceed 1 Mg ha^{-1} if cover crops are planted late in fall and terminated early in spring, which limits the period of time for cover crop growth (Table 2.1). Winter cover crops such as rye are predominantly used in cold temperate regions (Table 2.1). These cover crops are often added to continuous corn or rotations with corn such as corn–soybean and corn–small grains. Cover crop biomass production among cover crop species does not significantly differ although grass cover crops tend to produce more biomass than other species (Table 2.2).

2.2.2 Semiarid Temperate Regions

The amount of cover crop biomass produced in water-limited regions can be significant but is as variable as in regions with high precipitation. Cover crop biomass production in water-limited regions with <750 mm of annual precipitation can average 2.50 ± 2.37 Mg ha^{-1} (Blanco-Canqui et al., 2022; Ruis et al., 2019). This average across studies from semiarid regions suggests that cover crops can produce approximately 1 Mg ha^{-1} less biomass in water-limited regions relative to the average in regions with >750 mm of annual precipitation. Data on cover crop biomass production from arid regions (<250 mm precipitation) are unavailable. Low precipitation limits cover crop adoption in arid regions. As in high-precipitation regions, cover crop biomass production in low-precipitation regions is management dependent.

Table 2.2 Mean (± Standard Deviation) Cover Crop Biomass Production for Different Cover Crop Species for Two Climatic Regions Within Temperate Regions

Climate	Cover crop species	Cover crop biomass (Mg ha^{-1})
Humid temperate regions	Grasses	4.02 ± 3.47
	Legumes	3.58 ± 2.83
	Brassicas	2.73 ± 1.90
	Mixes	3.98 ± 2.83
Semiarid temperate regions	Grasses	3.37 ± 3.22
	Legumes	2.19 ± 1.94
	Brassicas	1.89 ± 1.22
	Mixes	2.66 ± 2.09

Notes: Ruis et al. (2019).

The same factors that affect cover crop biomass production in high-precipitation regions affect biomass production in low-precipitation regions. Grass cover crops and grass–legume mix cover crops tend to produce more biomass than legumes in semiarid temperate regions (Table 2.2). Also, winter (e.g., rye, triticale, wheat) and summer (e.g., sorghum sudangrass) cover crops can produce relatively high amounts of biomass relative to other species (Holman et al., 2018; Kumar et al., 2020). It is important to note that when cover crops produce significant amounts of biomass in water-limited regions, cover crops can reduce subsequent crop yields by reducing available water, especially in years when precipitation amount is below normal although improved soil physical properties after cover crop systems are introduced could lead to increased water retention in the long term (Holman et al., 2018; Nielsen et al., 2015). Improved cover crop management practices such as planting and/or terminating early can be options to reduce the negative impacts on subsequent crop yields. Well-established cover crops can increase water infiltration and reduce evaporation via residue mulching, which can be valuable in low-precipitation areas.

2.2.3 Tropical and Subtropical Regions

Cover crops generally produce greater amounts of biomass in subtropical and tropical regions than in temperate regions due to the longer growing window. For example, in a tropical region of Brazil, signal grass and jack bean produced approximately 9.32 Mg ha^{-1}, while sunn hemp, crowngrass, and pearl millet produced between 3.98 and 5.24 Mg ha^{-1} during the first year (Teixeira et al., 2014). Case studies from humid subtropical regions show cover crops can produce as much 16 Mg ha^{-1} of biomass with an average of 6 Mg ha^{-1} (Table 2.3). As in other regions, the amount of biomass can be variable, depending on cover crop species and management. Legume cover crops (e.g., clover, sunn hemp, hairy vetch) are often used in subtropical and tropical regions as N sources for subsequent crops. High humidity and temperature in subtropical and subtropical regions can lead to rapid decomposition of cover crops, which can positively or negatively impact soil ecosystem services. For instance, the rapid decomposition of cover crop residues

Table 2.3 Examples of Cover Crop Biomass Production Averaged Across Various Years in Subtropical and Tropical Climates

Location	Cover crop	Biomass (Mg ha^{-1})
[a]Parana, Brazil	Wheat	3.3
	Radish	3.7
	Vetch	3.7
	Lupine	3.8
	Oat	4.3
[b]Parana, Brazil	Vetch	3.2
	Radish	6.1
	Ryegrass	5.1
	Black oat	5.5
	Clover	2.2
[c]Georgia, U.S.	Rye	8.5
	Austrian winter pea	4.6
	Narrow-leaf lupine	6.7
	Cahaba vetch	2.8
	Crimson clover	3.2
[d]Rio Grande do Sul, Brazil	Oat	5.0
	Vetch	4.0
[e]Alabama, U.S.	Sunn hemp	7.6
[f]Georgia, U.S.	Rye	4.0
	Hairy vetch	4.2

[a] Tiecher et al. (2017).
[b] Pavinato et al. (2017).
[c] Webster et al. (2013).
[d] Gomes et al. (2009).
[e] Balkcom and Reeves (2005).
[f] Sainju et al. (2005).

can release nutrients for the next crop, but it may reduce soil C accumulation. In particular, legume cover crop residues are subject to rapid turnover relative to grass cover crop residues.

2.3 Belowground Biomass Production

Belowground cover crop biomass production can influence soil ecosystem services more than aboveground biomass because roots are in closer contact with the soil (Qi et al., 2022; Xu et al., 2021). The ecosystem service benefits of cover crop roots are many, but the question here is: How much root biomass do cover crops produce? Experimental data on cover crop root biomass production are fewer than on aboveground biomass production. The available data indicate cover crop root biomass production can be 40–50% of the aboveground biomass production. While cover crop root biomass production can be lower than aboveground biomass production, addition of cover crops to existing cropping systems increases the total amount of root biomass produced by the system. For instance, cover crops can add, on average, 1.25 Mg ha^{-1} of root biomass to cropping systems in water-limited regions considering the average aboveground cover crops biomass production in these regions is approximately 2.50 Mg ha^{-1} (Xu et al., 2021).

Similar to aboveground biomass production, cover crop root biomass production is highly variable due to differences in climate, length of cover crop growing season, cover crop species, and years after cover crop introduction. For example, just letting cover crops grow for a few days or weeks longer than early-terminated cover crops can significantly increase total root biomass production as well as aboveground biomass production. Selection of cover crop species can be important for increasing root biomass amount and the type of root biomass input. For example, species of brassicas have roots with larger diameter than grass and legume cover crops (Chen & Weil, 2010; Kemper et al., 2020). Differences in root biomass and characteristics among cover crops can determine differences in soil ecosystem services from cover crops.

How cover crop root biomass production varies with soil depth is unclear as most studies report root biomass data only for the upper 10 cm of soil depth. Cover crop roots that can penetrate deeper into the soil are important for deep C sequestration, nutrient scavenging, groundwater recharge, and other processes. In Germany, Kemper et al. (2020) found that root length density for seven select cover crop species was in this order: Oil radish, Winter turnip rape, and Phacelia > Bristle oat > Winter rye and Crimson clover. This order suggests that taprooted cover crops can penetrate deeper into the soil compared with fibrous-rooted cover crops. Planting a mix of both fibrous- and deep-rooted cover crops could be a strategy for a uniform root distribution in the soil profile.

Fibrous-rooted cover crops primarily improve soil properties near the soil surface, while deep-rooted cover crops can improve such properties in deeper depths. Overall, cover crops can produce significant amounts of root biomass (approximately 40–50% of aboveground biomass). Increasing the proportion of total cover crop biomass allocated to roots in deeper depths in the soil profile is critical to enhance soil ecosystem services.

2.4 Threshold Level of Biomass Production

Cover crops may not significantly improve soil ecosystem services unless biomass production exceeds a certain threshold level. The minimum level of cover crop biomass production needed to exert significant changes in soil ecosystem services could be approximately 1 Mg biomass ha^{-1} (Koehler-Cole et al., 2020; Kaspar et al., 2001; Finney & Kaye, 2017). Various field experiments have shown that cover crops, particularly winter cover crops, in cool temperate regions often produce less than 1 Mg ha^{-1} due to the short growing window and thus do not significantly alter soil ecosystem services (Table 2.1). The threshold level of biomass production can vary with the soil ecosystem service. For example, a larger amount of cover crop biomass may be needed to change soil C stocks and suppress weeds than for other soil services such as erosion control or nitrate leaching reduction (Koehler-Cole et al., 2020; Kaspar et al., 2001; Finney & Kaye, 2017). Also, the threshold level of cover crop biomass production to significantly suppress weeds can be approximately 4 Mg ha^{-1} (Finney & Kaye, 2017). Thus, developing management strategies that increase cover crop biomass production above the threshold level for different soils and climatic conditions is a priority to improve cover crop potential for maintaining or enhancing soil ecosystem services.

2.5 Management Practices that Affect Biomass Production

Cover crop biomass production not only depends on climate (e.g., precipitation and temperature) but also on cover crop management practices including planting and termination timing, seeding rates, cropping systems, irrigation, fertilization, and others (Table 2.4). In some cases, extending the cover crop growing season can increase cover crop biomass production if weather conditions are favorable. For instance, in cool temperate regions, winter cover crops such as winter rye are often planted late in fall after corn or soybean harvest and terminated early in spring before corn or soybean planting. Under this cover crop management scenario, the amount of biomass produced can be too low to exert significant

2.5 Management Practices that Affect Biomass Production

Table 2.4 Some of the Potential Strategies to Increase Cover Crop Biomass Production

Cover crop management strategies
1) Planting early via interseeding or drilling
2) Terminating late (at main crop planting)
3) Planting high-biomass species
4) Planting after corn silage or short-growing season crop
5) Increasing the seeding rate
6) Drilling rather than interseeding
7) Fertilizing at rates below main crop fertilization rates
8) Irrigating once or twice at establishment

Notes: Balkcom et al. (2018); Koehler-Cole et al. (2020); Ruis et al. (2017); Antosh et al. (2020).

Figure 2.2 Early-terminated (~30 days before planting corn) winter rye cover crop (left) can yield more biomass than late-terminated (at corn planting) winter rye cover crop (right) Photos by H. Blanco.

changes in soil ecosystem services (Table 2.1). Planting cover crops early and terminating early, planting late and terminating late, and combining early planting with late termination can be strategies to extend the growing season and increase cover crop biomass production (Figure 2.2).

2.5.1 Planting Early

2.5.1.1 Interseeding

Planting cover crops into standing crops, known as interseeding, can be a potential opportunity to increase the length of the cover crop growing season (Figure 2.3). Methods of cover crop interseeding include broadcasting, drilling, and broadcasting followed by drilling. Aerial interseeding with an airplane or drone is an option

Figure 2.3 Corn without cover crop interseeded (left) and corn with aerially interseeded approximately 30 days before corn harvesting (right; Blanco-Canqui et al., 2017) Photos by H. Blanco.

to interseed cover crops at large scales. Concerns exist, however, that interseeded cover crops can compete like weeds with main crops to reduce crop yields. Research shows that competition and reduction in crop yields due to interseeding can be minimal if cover crops are interseeded after main crops are well established. Specifically, cover crops interseeded after certain vegetative stages (e.g., V2 in corn) may not adversely affect yields. In Michigan, U.S., ryegrass cover crop broadcast interseeded with a hand-spreader at different corn growth stages from V2 to V7 for two years produced the highest amount of biomass of four treatments (ryegrass, crimson clover, oilseed radish, and three-species mix) and generally had no negative effects on subsequent crop yields (Brooker et al., 2020).

Interseeded cover crops may or may not produce more biomass than post-harvest planted cover crops (Table 2.1; Table 2.5). Limited seed–soil contact, seed trapping in crop leaves, and shading from crops can be some of the constraints that limit biomass production of broadcast interseeded cover crops. Broadcast interseeding with a high-clearance seeder equipped with nozzles that drop the seeds below the crop canopy can reduce seed trapping in the leaves, but seed–soil contact can still be limited. For example, studies in Michigan and Nebraska, U.S., found that broadcast interseeded grass cover crops produced variable amounts of biomass (0.2–3.4 Mg ha^{-1}; Blanco-Canqui et al., 2017; Brooker et al., 2020). Also, drill interseeded cover crops do not appear to yield more biomass than broadcast interseeded cover crops (Curran et al., 2018; Stanton & Haramoto, 2021). Overall, interseeded cover crops can produce similar amounts to cover crops seeded after harvest. Thus, extending the time between cover crop planting and termination via interseeding may not be a top strategy to increase biomass production.

Table 2.5 Examples of Cover Crop Biomass Production When Interseeded into Standing Crops in Late Summer

Location	Crop	Cover crop	Planting time before crop harvest	Years	Interseeding effect on cover crop biomass (Mg ha^{-1})
[a]Nebraska, U.S.	Corn	Winter rye	~ 1 month	3	<3.4
[b]Two sites in Minnesota, U.S.	Corn–soybean	Winter rye, red clover, hairy vetch, field pennycress, and mix (oat, pea, and tillage radish)	~ 4 months	3	< 1.6
[c]New York, Pennsylvania, and Maryland, U.S.	Corn	Annual ryegrass, red clover, crimson clover, and hairy vetch	3–4 months	1	< 2.8
[d]Three sites in Nebraska, U.S.	Soybean	Cereal rye and mix (rye, legume, and brassica)	~ 1 month	4	< 1.6
[e]Two sites in Michigan, U.S.	Corn	Ryegrass, crimson clover, oilseed radish, and mix	~ 4 (late) and 5 (early) months	2	< 0.2
[f]Minnesota, U.S.	Sugar beet	Austrian pea, winter camelina, brown mustard, and winter rye	~ 3 (early) and 2.6 (late) months	3	< 2.3

[a] Blanco-Canqui et al. (2017).
[b] Noland et al. (2018).
[c] Curran et al. (2018).
[d] Koehler-Cole et al. (2020).
[e] Brooker et al. (2020).
[f] Sigdel et al. (2021).

2.5.1.2 Planting after Summer Crop Harvest

Planting cover crops in summer after summer crop harvest (Table 2.6) can be a better option than either interseeding early (Table 2.5) or planting in late fall after crop harvest (Table 2.1). Data show that cover crops planted after crop harvest in summer often produce high amounts of biomass when soil moisture is adequate. Summer cover crops drilled after wheat harvest can produce as much as 8 Mg ha^{-1} in winter wheat-based rotations (Table 2.6). For instance, one of the dominant cropping systems in the U.S. Great Plains is winter wheat–fallow or winter wheat–summer crop–fallow. Replacing the fallow period with forage or cover crops in

Table 2.6 Some Examples of Cover Crop Biomass Production When Cover Crops are Seeded After Small Grains

Location	Crop	Cover crop growing season	Cover crop	Cover crop biomass (Mg ha^{-1})
[a]New Mexico, U.S.	Winter wheat	Spring	Canola	0.7b
			Oat	1.9a
			Pea	0.7b
[b]Kansas, U.S.	Winter wheat	Late fall to mid-spring	Hairy vetch	0.4d
		Late fall to mid-spring	Winter pea	0.6d
		Spring	Spring pea	1.6c
		Late fall	Winter triticale	4.1a
		Spring	Spring triticale	2.1b
[c]Manitoba, Canada	Spring wheat	Summer	Barley	7.1a
			Pea	7.8a
			Vetch	2.8b
[d]Kansas, U.S.	Winter wheat–grain sorghum	Mid-summer to late fall	Sunn hemp	7.4a
			Late-maturing soybean	6.0a

Means followed by different lowercase letters within each study are significantly different.
[a] Mesbah et al. (2019).
[b] Holman et al. (2018).
[c] Halde et al. (2014).
[d] Blanco-Canqui et al. (2011).

these cropping systems can be an opportunity to incorporate cover crops. The fallow period is intended to conserve water in water-limited regions, but research shows only approximately 30% of precipitation water is stored during the fallow period (Nielsen & Vigil, 2010).

Cropping systems with extended fallow periods often have lower amounts of crop residues, lower soil organic C levels, and higher rates of wind and wind erosion than systems with short fallow periods. Replacing fallow with cover crops can be feasible especially in years when precipitation is near normal. Grass cover crops replacing fallow in winter wheat–fallow systems in semiarid regions can produce over 4 Mg ha^{-1} of biomass (Table 2.6; Holman et al., 2018). In contrast, cover crops such as winter rye drilled in late fall and terminated early in spring under no-till continuous corn or corn–soybean systems often yield less than 2 Mg ha^{-1} due to less than optimum weather (e.g., low temperature) conditions for winter cover crop growth (Table 2.1). Planting cover crops in summer after small grain harvest has the potential to increase cover crop biomass production due to the longer growing window for cover crop growth after small grain wheat harvest relative to corn or soybean systems in cool temperate regions. However, the amount of cover crop biomass produced even when planted early after summer crop harvest varies, depending on soil moisture availability.

2.5.1.3 Planting after Corn Silage or Short-Growing Season Crop Harvest

Planting cover crops in late summer or early fall after harvest of corn silage, seed corn, and early-maturing crops or short-season hybrids can be another strategy to increase cover crop biomass production. For example, corn silage harvest leaves limited or no residue on the soil surface. Thus, cover crops in corn silage systems can establish well and produce significant amounts of biomass while protecting soil from erosion. The amount of cover crop biomass produced after corn silage harvest and terminated late the following spring (Table 2.7) can be 5- to 10-fold larger than that produced when the same cover crop is planted after grain corn harvest in late fall and terminated in early- or mid-spring (Table 2.1). If cover crops are planted after corn silage harvest in late summer but terminated the same year in fall or terminated early the following spring, cover crops could produce lower amounts of biomass than terminated late in spring (Anderson et al., 2022; Moore et al., 2014). Thus, even when cover crops are planted early under corn silage systems, termination time of cover crops determines the amount of cover crop biomass produced under these systems (Table 2.7). Data on cover crop biomass production after seed corn or after early-maturing crops are few, but these systems also allow extended time for cover crop growth similar to corn silage systems.

2.5.2 Terminating Late

Cover crops are commonly terminated several weeks before main crop planting to reduce risks of water depletion for the next crop or allow sufficient time for residue decomposition and nutrient release. This practice often results in low cover crop biomass production. Delaying cover crop termination for a few weeks or until main crop planting could be a strategy to enhance cover crop biomass production without reducing crop yields when precipitation is sufficient. Research shows that late termination of cover crops can more than double biomass production than early termination (Table 2.8; Figure 2.2). The increase in cover crop

Table 2.7 Examples of Cover Crop Biomass Production When Cover Crops are Seeded After Corn Silage Harvest

Location	Cover crop	Planting time	Termination time	Cover crop biomass (Mg ha^{-1})
[a]Nebraska, U.S.	Oat	Late summer	Late fall	1.9–2.2
[b]Nebraska, U.S.	Winter rye	Late summer	Late spring	8.2–12.1
[c]Pennsylvania, U.S.	Red clover	Late summer	Mid-spring	3.1
	Austrian winter pea			4.0
	Forage radish			2.3
	Canola			4.3
	Oat			3.6
	Rye			5.1
[d]Wisconsin, U.S.	Winter rye	Late summer	Early spring	4.0
[e]Iowa, U.S.	Winter rye	Late summer	Early spring	2.7
[f]Maryland, U.S.	Radish	Late summer	Early spring	2.6
	Winter rye			1.5
[g]Ohio, U.S.	Annual ryegrass	Late summer	Early spring	3.4–4.8

[a] Anderson et al. (2022).
[b] Blanco-Canqui et al. (2020).
[c] Hunter et al. (2019).
[d] Siller et al. (2016).
[e] Moore et al. (2014).
[f] White and Weil (2010).
[g] Fae et al. (2009).

biomass production with late termination under favorable weather conditions is large and consistent relative to early termination. Concerns, however, exist that delaying termination can reduce yields of subsequent crops due to water use and N immobilization. Residue decomposition for late-terminated cover crops such as grasses can be slower due to higher residue C to N ratio than for early-terminated cover crops. However, late-terminated cover crops may not always reduce subsequent crop yields (Otte et al., 2019). Delaying cover crop termination till planting is more likely to reduce subsequent yields in semiarid regions (<500 precipitation) due to limited precipitation input.

Table 2.8 Examples of How Termination Timing Affects Cover Crop Biomass Production

Location	Crop	Cover crop	Years in cover crop	Cover crop biomass (Mg ha^{-1})	
				Early termination	Late termination
[a]Nebraska, U.S.	Corn	Winter rye	3	0.3	1.9
[b]Pennsylvania, U.S.	Wheat, barley, and oat	Winter rye	4	1–7	up to 12
		Wheat		0.5–4	up to 9
[c]Maryland, U.S.	Corn	Winter rye	1	3.5	7.4
		Hairy vetch	1	1.2	4.9

[a] Ruis et al. (2017).
[b] Duiker (2014).
[c] Clark et al. (1994).

2.5.3 Cover Crop Mixes

It is often thought that multispecies cover crop mixes could produce more biomass than single species cover crops due to potential complementarity among diverse species. If this were the case, then cover crop mixes would improve soil ecosystem services more than single species. However, what does research indicate about biomass production of cover crop mixes? It indicates that when cover crop mixes are compared against all their constituents grown as single species, mixes do not generally produce more biomass than single species. A global review concluded that cover crops mixes yielded similar amount of biomass to single species in 88% of comparisons, yielded lower than high-biomass cover crops in 10% of comparisons, and yielded more in only 2% of comparisons although the latter comparison (2%) was confounded with the impacts of higher seeding rate for mixes than for single species (Florence & McGuire, 2020).

Based on the available research, increasing diversity of cover crops does not always translate into increased biomass production. High-biomass producing single species of cover crops can yield more than multispecies mixes in some cases (Finney & Kaye, 2017). Even reducing seeding rate for high-biomass cover crops and increasing seeding rate for low-biomass cover crops within a mix may not overcome the performance of high-biomass species. For example, grass cover crops such as winter rye when mixed with brassicas and legumes species tend to dominate the mix.

Note that some studies found greater biomass production for mixes than for single species, but such studies have compared mixes simply against one or none of the constituents grown as single species. This type of comparison is not valid as not all constituents in the mix were present as single species. In sum, experimental data indicate cover crop mixes do not outperform single species. Based on this finding, the amount of cover crop biomass produced can be more critical for improving soil ecosystem services than diversity or a combination of different cover crop species (Finney & Kaye, 2017; MacLaren et al., 2019).

Also note that it is important to mention that mixes can be advantageous over a single species when mixes provide soil cover at different times of the year, which is critical for N retention and weed suppression, among other benefits. For instance, in the northeastern U.S., forage radish–winter rye is a common cover crop mix (Wallace et al., 2021). In this scenario, radish often grows quickly in the fall and provides soil cover before winter rye, while winter rye grows in late winter or early spring and provides soil cover in spring. Thus, a mix of cover crop species that grow at different times can provide more soil and water conservation benefits compared with simply planting one species of cover crop.

2.5.4 Seeding Rate

Cover crop biomass production generally increases as cover crop seeding rate increases (Ruis et al., 2019). Identifying an optimum seeding rate for different cover crop species can be important to maximize cover crop production and minimize production costs. In a few cases, seeding rate may have a limited effect on cover crop biomass production. Haramoto (2019) reported that increasing seeding rate for winter rye from 34–112 kg seed ha^{-1} did not increase the total amount of biomass produced. The impact of increasing seeding rates on biomass production can depend on cover crop species and cover crop growth stage. Increasing seeding rates can particularly enhance cover crop biomass production during the early and middle growth stages of cover crops but may have limited effects at later stages. For instance, Boyd et al. (2009) found increasing winter rye seeding rates (90, 180, and 270 kg ha^{-1}) increased biomass production up to 70 or 80 days after planting but not in later stages of growth. In general, cover crop biomass production increases with an increase in cover crop seeding rate in most cases.

2.5.5 Planting Method

The method used to plant cover crops can affect seed placement, germination, and thus biomass production. Two common methods for planting cover crops include drilling and broadcasting. Drilling is more common than broadcasting as it ensures adequate seed–soil contact. Broadcasting with a high-clearance equipment and aerial broadcasting with an airplane are two methods that allow cover crop planting before main crop harvest. Broadcasting often leaves seeds on the soil surface with limited contact with the soil and soil water, which can reduce seed germination and establishment. However, it can allow planting large fields in a shorter amount of time than the drill-planting method.

A review reported cover crop biomass production was 3.39 ± 2.93 Mg ha^{-1} for drilling and 2.56 ± 2.16 Mg ha^{-1} for broadcasting, indicating drilling can produce more biomass than broadcasting due to better cover crop establishment (Ruis et al., 2019). Because cover crop germination can be lower under broadcasting, higher seeding rates may be needed for broadcasting than for drilling to produce similar amounts of biomass. Drilling is favored over broadcasting, particularly in relatively dry soils, to facilitate cover crop establishment. Also, the impact of planting method on cover crop biomass production can depend on seeding rate and cover crop species. In South Carolina, under the same seeding rate, drill-planted crimson clover produced more biomass than drill-planted hairy vetch and that hairy vetch produced more biomass under broadcasting (St. Aime et al., 2021). Overall, cover crop biomass production can be greater under drill-planted than broadcast-planted cover crops when all other factors are equal.

2.5.6 Tillage and Cropping Systems

Tillage systems are often classified as no-till, reduced till, and conventional till systems. Under no-till, cover crop residues are left on the soil surface after termination while, under reduced and conventional till, cover crop residues are incorporated into the soil as green manure. It is often considered that cover crops and no-till can have synergistic effects on cover crop biomass production, but experimental data indicate cover crops managed under no-till do not normally produce more biomass than those managed under reduced and conventional till. A review of studies across temperate regions found that cover crops produced 3.27 ± 3.04 Mg ha^{-1} under tilled and 3.13 ± 2.43 Mg ha^{-1} under no-till management (Ruis et al., 2019). In some cases, tilled soils could enhance cover crop biomass production due to higher soil temperature and better germination when compared against no-till systems especially in cold and clayey soils. However, pairing cover crops with no-till management should still be favored to better protect soil from erosion and potentially lower losses of C compared with cover crops combined with tilled systems. Overall, cover crop biomass production is not largely affected by the type of tillage system.

Cropping systems could have greater impact on cover crop biomass than tillage systems. Cover crops can produce more biomass following low residue-producing crops such as soybean and vegetables than following high-biomass crops such as corn. For example, in Italy, hairy vetch produced 7.2 Mg ha^{-1} of biomass and oat produced 5.1 Mg ha^{-1} when planted as winter cover crops after tomato crops (Campiglia et al., 2010). In Canada, oilseed radish produced 6.1 Mg ha^{-1} of biomass and rye produced 1.6 Mg ha^{-1} under tomato crops (Chahal & Van Eerd, 2020).

Additionally, cover crops can produce more biomass in fields where crop residues are partly or completely removed for expanded uses (e.g., livestock production) than in fields with no removal (Anderson et al., 2022). Cover crop biomass production under soils with limited surface cover can be greater due to better seed–soil contact and warmer soil temperature in spring, which can enhance cover crop germination and growth relative to soils with abundant residue cover. Evaporation can be high and soil water content can be low under low residue cover, but when precipitation is abundant, cover crop can produce significant amounts of biomass under low-biomass crops or when crop residues are removed.

2.5.7 Soil Texture and Number of Years in Cover Crops

It is hypothesized that soil texture can indirectly affect cover crop biomass production by influencing soil processes including water retention, nutrient availability, soil compaction, aeration, thermal conductivity, and others. For instance, medium-textured soils can hold more water and nutrients than coarse-textured soils, and potentially better support plant growth. However, studies specifically correlating soil particle size distribution against cover crop biomass production are unavailable. Unlike water and nutrient availability, which directly affect plant growth, soil particle size distribution effects on plant growth appear to be complex. Experimental data across a soil texture gradient under the same cover crop management and climatic conditions can be valuable to further our understanding of how an increase in sand or clay content affects biomass production.

Similar to soil texture, the number of years after cover crop introduction appears to have small impacts on cover crop biomass. Annual fluctuations in precipitation amount and cover crop growing window due to differences in planting and termination dates have larger impact than years under cover crops. A review found that aboveground biomass production did not significantly change with time after cover crop adoption (Ruis et al., 2019). The same review, however, found root biomass accrual tended to differ with time after cover crop introduction. Root biomass production can be generally greater in the second than in the first year, likely due to better establishment of cover crops with time after adoption and potential improvement in soil properties with time after cover crop establishment. In conclusion, soil texture and time after cover crop introduction do not seem to have a large influence on cover crop production.

2.5.8 Irrigation and Fertilization

Irrigating or fertilizing cover crops can maximize cover crop biomass production and potentially further improve soil ecosystem services (Table 2.4). Irrigation is often used in water-limited regions to successfully establish cover crops as simply planting cover crops in these regions does not guarantee that cover crop will produce adequate amounts of biomass. A study comparing biomass production between rainfed and irrigated cover crops found that flax, oat, pea, rapeseed, and a 10-species mixture cover crops produced, on average, 3.02 Mg ha^{-1} under rainfed conditions and 4.27 Mg ha^{-1} under irrigated conditions across two years and two sites in the central Great Plains (Nielsen et al., 2015). These data show that irrigation can increase cover crop production over non-irrigated cover crops. The positive impact of irrigation on biomass production in water-limited regions can become larger compared to rainfed cover crop as precipitation amount decreases. In some cases, one or two irrigation events may be sufficient to establish cover crops and produce significant amounts of biomass. In the southwestern U.S., cover crops irrigated once (38 mm water) at establishment produced as much as 2.3 Mg ha^{-1} (Ghimire et al., 2019). More frequent irrigation can further increase cover crop biomass production (Antosh et al., 2020; Nielsen et al., 2015).

Similarly, low fertility soils or soils with low residual N could benefit from fertilization of non-legume cover crops to produce an adequate amount of biomass. Balkcom et al. (2018) found that winter rye cover crop biomass production linearly increased as the N application rate gradually increased from 0 to 101 kg N ha^{-1}. The addition of 101 kg ha^{-1} nearly tripled rye cover crop biomass production (6 Mg ha^{-1}) relative to no N addition (2.2 Mg ha^{-1}). While the relationship between cover crop biomass production and N application rates can be linear at low N rates, such relationship is expected to follow a quadratic pattern with increasing rates of N application. Because fertilization of cover crops is uncommon, no current recommendation on the amount of N required for cover crops exists although the N requirement will most probably be lower than for main crops.

Adding animal manure can be an alternative to inorganic fertilization, depending on availability and cost. In some cases, inorganic fertilizer combined with animal manure can be a better option than either inorganic fertilizer or animal manure alone. While irrigation and fertilization can result in a significant increase in biomass production, the economic impacts of such practices need, however, consideration. Grazing cover crops and improvement in soil ecosystem services could justify the use of fertilizers and irrigation to maximize cover crop production. Irrigation and fertilization can be used as a strategy to achieve a threshold level of biomass needed to improve specific soil ecosystem services if non-fertilized or non-irrigated cover crops do not produce sufficient biomass.

2.6 Summary

Literature indicates that cover crops can produce 3.78 ± 3.08 Mg ha^{-1} across all temperate regions and 2.40 ± 2.33 Mg ha^{-1} in semiarid regions. Cover crops often produce larger amounts of biomass in subtropical and tropical regions than in temperate regions, but residue decomposition rates are lower in temperate regions. Belowground cover crop biomass production can be 40–50% of aboveground biomass production but data on belowground biomass production are fewer than on aboveground biomass production.

Cover crop biomass production is highly variable as a function of climate and cover crop planting and termination dates, seeding rate, growing season, and planting method, among others. Cover crops produce less biomass in colder than in warmer regions due to shorter growing cover crop season in colder regions. Cover crop mixes and single species produce similar amounts of biomass. Indeed, high-biomass single species of cover crops could produce more biomass than mixes.

Increasing the cover crop growing window by planting cover crops early and terminating them late can be strategies to increase biomass production. Additionally, irrigation and fertilization can increase cover crop biomass production although the increased biomass production costs need consideration. Drilling of cover crops ensures better seed–soil contact and can thus lead to greater biomass production than broadcast-planted cover crops. Tillage system, soil texture, cover crop mixes, and time after cover crop introduction do not appear to have large effects on biomass production relative to cover crop growing window. Overall, cover crops can produce significant amounts of biomass, but such production can be highly variable among and within locations.

References

Anderson, L., Blanco-Canqui, H., Drewnoski, M., & MacDonald, J. (2022). Cover crop grazing impacts on soil properties and crop yields under irrigated no-till corn-soybean. *Soil Science Society of America Journal, 86*, 118–133.

Antosh, E., Idowu, J., Schutte, B., & Lehnhoff, E. (2020). Winter cover crops effects on soil properties and sweet corn yield in semi-arid irrigated systems. *Agronomy Journal, 112*, 92–106.

Balkcom, K. S., Duzy, L. M., Arriaga, F. J., Delaney, D. P., & Watts, D. B. (2018). Fertilizer management for a rye cover crop to enhance biomass production. *Agronomy Journal, 110*, 1233–1242.

Balkcom, K. S., & Reeves, D. W. (2005). Sunn-hemp utilized as a legume cover crop for corn production. *Agronomy Journal, 97*, 26–31.

Blanco-Canqui, H., Drewnoski, M., Redfearn, D., Parsons, J., Lesoing, G., & Tyler, W. (2020). Does cover crop grazing damage soils and reduce crop yields? *Agrosystems, Geosciences and Environment, 3*, e20102.

Blanco-Canqui, H., Mikha, M., Presley, D. R., & Claassen, M. M. (2011). Addition of cover crops enhances no-till potential for improving soil physical properties. *Soil Science Society of America Journal, 75*, 1471–1482.

Blanco-Canqui, H., Ruis, S., Holman, H., Creech, C., & Obour, A. (2022). Can cover crops improve soil ecosystem services in water-limited environments? A review. *Soil Science Society of America Journal, 86*, 1–18.

Blanco-Canqui, H., Sindelar, M., & Wortmann, C. S. (2017). Aerial interseeded cover crop and corn residue harvest: Soil and crop impacts. *Agronomy Journal, 109*, 1344–1351.

Boyd, N. S., Brennan, E. B., Smith, R. F., & Yokota, R. (2009). Effect of seeding rate and planting arrangement on rye cover crop and weed growth. *Agronomy Journal, 101*, 47–51.

Brooker, A. P., Renner, K. A., & Sprague, C. L. (2020). Interseeding cover crops in corn. *Agronomy Journal, 112*, 139–147.

Campiglia, E., Caporali, F., Radicetti, E., & Mancinelli, R. (2010). Hairy vetch (*Vicia villosa* Roth.) cover crop residue management for improving weed control and yield in no-tillage tomato (*Lycopersicon esculentum* mill.) production. *European Journal of Agronomy, 33*, 94–102.

Chahal, I., & Van Eerd, L. L. (2020). Cover crop and crop residue removal effects on temporal dynamics of soil carbon and nitrogen in a temperate, humid climate. *PLoS One, 15*. http://dx.doi.org/10.1371/journal

Chen, G., & Weil, R. R. (2010). Penetration of cover crop roots through compacted soils. *Plant and Soil, 331*, 31–43.

Clark, A. J., Decker, A. M., & Meisinger, J. J. (1994). Seeding rate and kill date effects on hairy vetch-cereal rye cover crop mixtures for corn production. *Agronomy Journal, 86*, 1065–1070.

Curran, W. S., Hoover, R. J., Roth, G. W., Wallace, J. M., Dempsey, M. A., Mirsky, S. B., Ackroyd, V. J., Ryan, M. R., & Pelzer, C. J. (2018). Evaluation of cover crops drill-interseeded into corn across the mid-Atlantic. *Crops Soils, 51*, 18.

Duiker, S. W. (2014). Establishment and termination dates affect fall-established cover crops. *Agronomy Journal, 106*, 670–678.

Fae, G. S., Sulc, R. M., Barker, D. J., Dick, R. P., Eastridge, M. L., & Lorenz, N. (2009). Integrating winter annual forages into a no-till corn silage system. *Agronomy Journal, 101*, 1286–1296.

Finney, D. M., & Kaye, J. P. (2017). Functional diversity in cover crop polycultures increases multifunctionality of an agricultural system. *Journal of Applied Ecology, 54*, 509–517.

Florence, A. M., & McGuire, A. M. (2020). Do diverse cover crop mixtures perform better than monocultures? A systematic review. *Agronomy Journal, 112*, 3513–3534.

Ghimire, R., Ghimire, B., Mesbah, A. O., Sainju, U. M., & Idowu, O. J. (2019). Soil health responses of cover crops in winter wheat-fallow system. *Agronomy Journal, 111*, 2108–2115.

Gomes, J., Bayer, C., Costa, F. S., Piccolo, M. C., Zanatta, J. A., Vieira, F. C. B., & Six, J. (2009). Soil nitrous oxide emissions in long-term cover crops-based under subtropical climate. *Soil and Tillage Research, 106*, 36–44.

Halde, C., Gulden, R. H., & Entz, M. H. (2014). Selecting cover crop mulches for organic rotational no-till systems in Manitoba. *Canadian Agronomy Journal, 106*, 1193–1204.

Haramoto, E. R. (2019). Species, seeding rate, and planting method influence cover crop services prior to soybean. *Agronomy Journal, 111*, 1068.

Holman, J. D., Arnet, K., Dille, J., Maxwell, S., Obour, A., Roberts, T., Roozeboom, K., & Schlegel, A. (2018). Can cover or forage crops replace fallow in the semiarid central Great Plains? *Crop Science, 58*, 932–944.

Hunter, M. C., Schipanski, M. E., Burgess, M. H., LaChance, J. C., Bradley, B. A., Barbercheck, M. E., & Mortensen, D. A. (2019). Cover crop mixture effects on maize, soybean, and wheat yield in rotation. *Agricultural and Environmental Letters, 4*, 1–5.

Kaspar, T. C., Radke, J. K., & Laflen, J. M. (2001). Small grain cover crops and wheel traffic effects on infiltration, runoff, and erosion. *Journal of Soil and Water Conservation, 56*, 160–164.

Kemper, R., Bublitz, T. A., Müller, P., Kautz, T., Döring, T. F., & Athmann, M. (2020). Vertical root distribution of different cover crops determined with the profile wall method. *Agriculture, 10*, 503.

Koehler-Cole, K., Elmore, R. W., Blanco-Canqui, H., Francis, C. A., Shapiro, C. A., Proctor, C. A., Heeren, D. M., Ruis, S., Irmak, S., & Ferguson, R. B. (2020). Cover crop productivity and subsequent soybean yield in the western Corn Belt. *Agronomy Journal, 112*, 2649–2663.

Kumar, A. V., Obour, A., Jha, P., Manuchehri, M. R., Dille, J. A., Holman, J., & Stahlman, P. W. (2020). Integrating cover crops for weed management in the semi-arid U.S. Great Plains: opportunities and challenges. *Weed Science, 68*, 311–323.

MacLaren, C., Swanepoel, P., Bennett, J., Wright, J., & Dehnen-Schmutz, K. (2019). Cover crop biomass production is more important than diversity for weed suppression. *Crop Science, 59*, 733–748.

Mesbah, A., Nilahyane, A., Ghimire, B., Beck, L., & Ghimire, R. (2019). Efficacy of cover crops on weed suppression, wheat yield, and water conservation in winter wheat-sorghum-fallow. *Crop Science, 59*, 1745–1752.

Moore, E. B., Wiedenhoeft, M. H., Kaspar, T. C., & Cambardella, C. A. (2014). Rye cover crop effects on soil quality in no-till corn silage–soybean cropping systems. *Soil Science Society of America Journal, 78*, 968–976.

Nielsen, D. C., Lyon, D. J., Hergert, G. W., Higgins, R. K., & Holman, J. D. (2015). Cover crop biomass production and water use in the central Great Plains. *Agronomy Journal, 107*, 2047–2058.

Nielsen, D. C., & Vigil, M. F. (2010). Precipitation storage efficiency during fallow in wheat-fallow systems. *Agronomy Journal, 102*, 537–543.

Noland, R. L., Wells, M. S., Sheaffer, C. C., Baker, J. M., Martinson, K. L., & Coulter, J. A. (2018). Establishment and function of cover crops interseeded into corn. *Crop Science, 58*, 863–873.

Otte, B., Mirsky, S., Schomberg, H., Davis, B., & Tully, K. (2019). Effect of cover crop termination timing on pools and fluxes of inorganic nitrogen in no-till corn. *Agronomy Journal, 111*(6), 2832–2842.

Pavinato, P. S., Rodrigues, M., Soltangheisi, A., Sartor, L. R., & Withers, P. J. A. (2017). Effects of cover crops and phosphorus sources on maize yield, phosphorus uptake, and phosphorus use efficiency. *Agronomy Journal, 109*, 1039–1047.

Qi, J., Jensen, J. L., Christensen, B. T., & Munkholm, L. J. (2022). Soil structural stability following decades of straw incorporation and use of ryegrass cover crops. *Geoderma, 406*, 115463.

Rankoth, L. M., Udawatta, R. P., Gantzer, C. J., Jose, S., & Nelson, K. A. (2019). Cover crop effects on corn plant sap flow rates and soil water dynamics. *Crop Science, 59*, 227–2236.

Ruis, S. J., Blanco-Canqui, H., Creech, C. F., Koehler-Cole, K., Elmore, R. W., & Francis, C. A. (2019). Cover crop biomass production in temperate agroecozones. *Agronomy Journal, 111*, 1535–1551.

Ruis, S. J., Blanco-Canqui, H., Jasa, P. J., Ferguson, R. B., & Slater, G. (2017). Can cover crop use allow increased levels of corn residue removal for biofuel in irrigated and rainfed systems? *Bioenergy Research, 10*, 992–1004.

Sainju, U. M., Whitehead, W. F., & Singh, B. P. (2005). Biculture legume-cereal cover crops for enhanced biomass yield and carbon and nitrogen. *Agronomy Journal, 97*, 1402–1412.

Sigdel, S., Chatterjee, A., Berti, M., Wick, A., & Gasch, C. (2021). Interseeding cover crops in sugar beet. *Field Crops Research, 263*, 1080179.

Siller, A. R. S., Albrecht, K. A., & Jokela, W. E. (2016). Soil erosion and nutrient runoff in corn silage production with kura clover living mulch and winter rye. *Agronomy Journal, 108*, 989–999.

Sindelar, M., Blanco-Canqui, H., Virginia, J., & Ferguson, R. (2019). Cover crops and corn residue removal: Impacts on soil hydraulic properties and their relationships with carbon. *Soil Science Society of America Journal, 83*, 221–231.

St. Aime, R., Noh, E., Jr. Bridges, W. C., & Narayanan, S. A. (2021). Comparison of drill and broadcast planting methods for biomass production of two legume cover crops. *Agronomy, 12*, 79.

Stanton, V. L., & Haramoto, E. R. (2021). Biomass potential of drill interseeded cover crops in corn in Kentucky. *Agronomy Journal, 113*, 1238–1247.

Teixeira, R. A., Soares, T. G., Fernandes, A. R., & Braz, A. M. S. (2014). Grasses and legumes as cover crop in no-tillage system in northeastern Pará Brazil. *Acta Amaz, 44*, 411–418.

Tiecher, T., Calegari, A., Caner, L., & Rheinheimer, D. D. S. (2017). Soil fertility and nutrient budget after 23-years of different soil tillage systems and winter cover crops in a subtropical Oxisol. *Geoderma, 308*, 78–85.

Wallace, J., Isbell, S., Hoover, R., Barbercheck, M., Kaye, J., & Curran, W. (2021). Drill and broadcast establishment methods influence interseeded cover crop performance in organic corn. *Renewable Agriculture and Food Systems, 36*, 77–85.

Webster, T. M., Scully, B. T., Grey, T. L., & Culpepper, A. S. (2013). Winter cover crops influence Amaranthus palmeri establishment. *Crop Protection, 52*, 130–135.

White, C. M., & Weil, R. R. (2010). Forage radish and cereal rye cover crop effects on mycorrhizal fungus colonization of maize roots. *Plant and Soil, 328*, 507–521.

Xu, H., Vandecasteele, B., De Neve, S., Boeckx, P., & Sleutel, S. (2021). Contribution of above- versus belowground C inputs of maize to soil organic carbon: Conclusions from a 13C/12C-resolved resampling campaign of Belgian croplands after two decades. *Geoderma, 383*, 114727.

3

Soil Health

3.1 Soil Health

Soil health refers to the ability of the soil to function as a living system that supports plant, animal, human, and overall ecosystem health (Doran & Zeiss, 2000; Lehmann et al., 2020). It considers soil as a living entity, similar to plants, animals, and humans. Soil health is a holistic concept with broad implications for planetary health in the face of increasing climatic fluctuations, food insecurity, and soil and environmental degradation. Soil health is emerging as the centerpiece to address the contemporary agronomic, environmental, soil, and economic challenges (van Bruggen et al., 2019). Soil health is a concept designed to develop our understanding of the close interrelationship of soil with plants, animals, and humans.

Embracing soil health is not only about producing crops or ensuring food security but also about mitigating and adapting to increasing climatic fluctuations, improving water and air quality, restoring degraded soil ecosystem services, and preventing pests and diseases, among others. Indeed, soil health is vital to the delivery of soil ecosystem services. In other words, a healthy soil should deliver all the essential soil ecosystem services (Figure 3.1). A healthy soil should be able to absorb and hold water during droughts, release water during floods, moderate abrupt fluctuations in soil temperature, sequester C from the atmosphere via plants, filter pollutants from water, retain and provide nutrients, support biological communities, and perform other vital functions. Restoring, maintaining, and improving the health of soils are essential prerequisites for the continued delivery of ecosystem services from soils.

Soil health is related to the soil quality concept, but soil health has much broader connotations. Soil quality refers to the capacity of the soil to perform a specific

---

Cover Crops and Soil Ecosystem Services, First Edition. Humberto Blanco.
© 2023 American Society of Agronomy, Inc. / Crop Science Society of America, Inc.
/ Soil Science Society of America, Inc. Published 2023 by John Wiley & Sons, Inc.

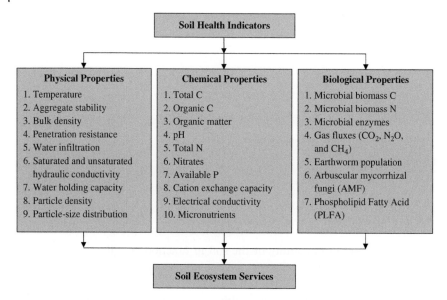

Figure 3.1 Changes in soil physical, chemical, and biological properties as indicators of soil health determine the extent of ecosystem services from cover crops.

function, while the former considers soil as a living and dynamic system with multiple social, economic, and political dimensions. Soil health is an emerging paradigm with measurement standards (e.g., scores, indexes) still under development. While some may consider this concept as subjective or qualitative, the quantification of changes in soil physical, chemical, and biological properties indicates changes in soil health. The changes in soil physical, chemical, and biological properties and their interconnectedness determine soil health and the ecosystem services that soils provide (Figure 3.1).

Most soil properties are dynamic and can rapidly change with management shifts, especially near the soil surface. Among the dynamic soil properties are soil water content, soil temperature, soil sorptivity (initial water infiltration), aggregate stability, particulate or labile C, N fractions, microbial biomass, and earthworm abundance (Figure 3.1). Advanced techniques for the characterization of changes in soil health and development of standard indexes are needed to fully characterize soil health. Remote sensing via satellites and drones, robotics, and other field approaches are upcoming tools for rapid and accurate characterization of soil properties, which can further our understanding of how different management strategies such as cover crops alter soil health at different temporal and spatial scales.

3.2 Cover Crops and Soil Health

Introduction of cover crops could enhance soil ecosystem services if cover crops improve the health of the soil as characterized by changes in physical, chemical, and biological properties. For instance, a potential improvement in soil aggregate stability, macroporosity, and hydraulic properties with cover crops can directly and indirectly affect C sequestration, nutrient cycling (e.g., retention, release), and water quality (e.g., filtration of pollutants), soil biological activity, and other soil ecosystem services. Also, an increase in soil organic matter concentration and available water can affect soil resilience against droughts.

It is commonly considered that introduction of cover crops would readily improve soil physical, chemical, and biological properties. However, what do research data show about the impacts of cover crops on these soil properties? Do cover crops improve all soil properties? What are the factors that may affect cover crop impacts? A discussion based on experimental data is needed to advance our understanding of how cover crop introduction affects soil physical, chemical, and biological properties as indicators of soil health and thus soil ecosystem services.

3.2.1 Soil Physical Properties

Soil physical properties are key determinants of water, air, and heat transmission in the soil. Experimental data indicate that cover crop impacts on soil physical properties can be site-specific and biomass production driven. Cover crops generally reduce soil penetration resistance; increase wet aggregate stability, macroporosity, water infiltration, and air permeability; and moderate soil temperature (Table 3.1). However, cover crops can have mixed effects on other physical properties including soil bulk density, dry aggregate stability, hydraulic conductivity, water content at field capacity and permanent wilting point, and plant available water (Irmak et al., 2018). A site-specific evaluation of all soil physical properties is recommended to assess whether cover crops are affecting all or only some physical properties. Changes in soil physical properties observed in one field may not be observed in the neighboring field due to differences in cover crop management, soil type, and other factors. Also, cover crop effects on soil physical properties develop over several years and are more often measurable in the long term compared to one to three years (Steele et al., 2012).

3.2.1.1 Soil Compaction

Soil compaction under highly mechanized agriculture is an increasing concern and threat to soil health. Research shows cover crops can generally reduce soil penetration resistance, which is an indicator of soil compaction (Table 3.1). Soil penetration resistance simulates the resistance of soil to root penetration.

Table 3.1 Review Results on the Impacts of Cover Crop on Soil Physical Properties from 98 studies

Soil physical property	Cover crop effect	Change
Penetration resistance	Reduced	5%
Wet aggregate stability	Increased	16%
Amount of macropores	Increased	1.5%
Water infiltration	Increased	62%
Air permeability	Increased	132%
Daytime temperature	Reduced	−2 °C
Nighttime temperature	Increased	1 °C
Other physical properties	Mixed or no effect	Variable

Notes: Blanco-Canqui & Ruis (2020).

The decrease in penetration resistance with cover crops can have important implications for the management of soil compaction. Reduced soil compaction via cover crops can directly lead to increased crop yields relative to fields without cover crops (Zhang et al., 2022). An increase of soil penetration resistance values to greater than 2 MPa in the root zone can reduce root growth, water and nutrient accessibility, and overall soil health although the exact threshold value of penetration resistance depends on crop and soil type (Colombi et al., 2018).

Introducing cover crops into a cropping system can reduce risks of soil compaction by increasing soil aggregation, macroporosity, and soil organic matter concentration (Table 3.1). For instance, an increase in soil organic matter with cover crops can reduce soil compactibility (e.g., susceptibility of the soil to compaction) or compressibility by increasing the rebounding capacity and elasticity of the soil (Blanco-Canqui et al., 2011; Díaz-Zorita & Grosso, 2000). Further, soils with cover crops could allow animal or machine traffic at higher soil water contents with no major risks of soil compaction than soils without cover crops (Blanco-Canqui et al., 2011). Thus, if soil organic matter concentration increases with cover crops, one might be able to perform field operations earlier in spring when soils are often wet without significant concerns of soil compaction (Blanco-Canqui et al., 2011; Díaz-Zorita & Grosso, 2000).

Furthermore, cover crops may not only reduce soil compaction by improving related soil properties such as the amount of organic matter, aggregation, and biological activity (e.g., fungal hyphae) but also by directly bio-drilling the soil. Tap-rooted cover crops such as forage radish can penetrate compacted soil layers and alleviate soil compaction (Chen & Weil, 2011; Zhang et al., 2022). Bio-drilling also creates large root channels, which can allow water accumulation from rain or

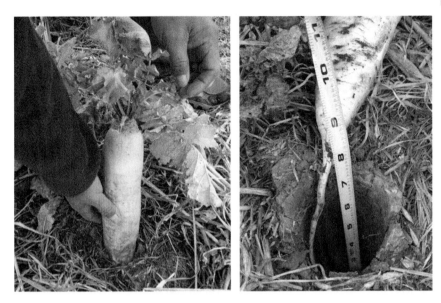

Figure 3.2 Tap-rooted cover crops can penetrate 20 cm or deeper into the soil and break up compacted layers. Photo by H. Blanco.

irrigation, thereby reducing runoff, increasing soil water content, and potentially increasing crop yields (Figure 3.2).

Subsoiling and deep tillage are traditional strategies to break compacted layers, but these practices can be both costly and a temporary solution for compaction (Zhang et al., 2022). Tilled soils are susceptible to rapid consolidation or recompaction. Introducing cover crops can be a biological way of building soil resistance and resilience against soil compaction risks. If bio-tillage is desired to open up compacted soil layers, planting tap-rooted cover crops can be a strategy. Finally, combining cover crops with controlled traffic farming, which confines traffic to permanent lanes within a field, and diversified crop rotations is recommended as an option to manage excessive soil compaction rather than simply relying on cover crops alone.

3.2.1.2 Soil Structure

Changes in soil aggregate stability and pore size distribution are some of the indicators of changes in soil structural health. Research shows cover crops can increase wet aggregate stability more rapidly than other soil properties although most of the increases occur near the soil surface (<10 cm depth; Table 3.1; Nouri et al., 2019). An increase in soil aggregate stability increases macroporosity and reduces microporosity. Cover crops can thus alter the distribution of pore sizes in the soil (Abdollahi et al., 2014). Macropores, (>500 mm) are essential for the rapid exchange of water, heat, and gas through the soil, but some proportion of mesopores, (5–500 mm), and micropores, (<5 mm) are also important (Singh et al., 2020). Micropores can improve water retention at high negative matric potentials. For example, plant available water is retained inside mesopores and micropores as water in the macropores drains rapidly under gravity.

Cover crops can maintain or improve near-surface wet aggregate stability and pore size by (a) providing surface cover, (b) increasing root biomass input, (c) increasing soil organic matter concentration, and (d) improving soil biological properties (Abdollahi et al., 2014). First, cover crops via canopy cover or residues protect soil aggregates near the soil surface from the erosive forces of raindrops that cause aggregate breakdown and dispersion (Singh et al., 2020). The effectiveness of cover crops to protect soil aggregates depends, however, on the density of cover crop canopy as well as the amount of biomass produced. Second, cover crop roots can be critical to the formation and stability of soil aggregates. Roots not only release organic binding agents for aggregate formation but also enmesh soil particles into aggregates. Third, an increase in wet aggregate stability and pore size is related to changes in soil organic matter concentration with cover crops. Soil structural properties are strongly correlated with an increase in soil organic matter concentration (Blanco-Canqui et al., 2011). Fourth, cover crop biomass provides an energy source to microorganisms, which can release polysaccharides and other organic compounds to promote soil aggregation (Qi et al., 2021).

It is important to note that cover crop-induced changes in wet aggregate stability, similar to other soil dynamic properties, can vary among cover crop growing seasons and even within the same season. Cover crop effects on soil structural properties are often the largest when cover crops are growing but diminish after cover crop termination. Assessment of soil aggregate stability and other dynamic properties on a temporal scale within the same year can provide a better understanding of how dynamic soil properties, thus surface soil health, respond to cover crop management with time. Also, planting cover crops after each main crop harvest is important to maintain soil benefits from cover crops (Blanco-Canqui & Ruis, 2020). For example, soil aggregation near the soil surface can rapidly decline after cover crop termination, especially when cover crop residue amount is limited (Figure 3.3). Overall, cover crops generally improve wet aggregate stability, but their impacts on dry aggregate stability seem to be limited.

3.2.1.3 Water Infiltration

Changes in water infiltration determine the amount of rain, snowmelt, and irrigation water that enters the soil. Cover crops can increase water infiltration by approximately 60% in most cases (Table 3.1). Improvement in soil aggregation and macroporosity, reduction in surface sealing, development of biological channels

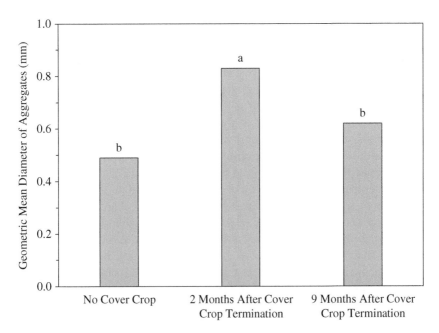

Figure 3.3 Winter triticale cover crop effect on geometric mean diameter of water-stable aggregates after two and nine months following cover crop termination in a semiarid soil in the U.S. Great Plains. Bars with the same letter are not statistically different. (Blanco-Canqui et al., 2013).

(e.g., earthworm burrows), and preservation of surface-exposed macropores after cover crop adoption are some of the mechanisms by which cover crops can contribute to increased water infiltration (Zhang et al., 2022). Deep and extensive roots upon decomposition can also develop conduits for water to infiltrate into deeper soil layers (Figure 3.2). Water infiltration may not change unless the proportion of macropores in the soil increases after cover crop adoption (Irmak et al., 2018).

The increased water infiltration with the introduction of cover crops can improve the capacity of the soil to absorb and capture water. Water infiltration determines rainfall or irrigation water partitioning into runoff. In sloping fields, the higher the water infiltration, the lower the amount of rainfall or irrigation water available for runoff (Kaspar et al., 2001). In nearly-level fields, the higher the water infiltration, the lower the risks for water ponding and flooding under intense rainstorms (Kahimba et al., 2008). In water-limited regions, an increase in infiltration can increase water storage in the soil and reduce the negative effects of cover crops on water depletion. The increased water infiltration, as reported in most cover crop studies, can thus have important implications for the management of soil water in agricultural lands.

3.2.1.4 Temperature

Soil temperature is highly sensitive to changes in surface cover. Canopy cover and residue input from cover crops can rapidly alter soil temperature by insulating the soil surface and reflecting the solar radiation. Cover crop effects on soil temperature can be larger in soils with limited crop residue cover than in those with abundant crop residue cover. Changes in soil temperature can be measurable shortly after cover crop establishment due to surface shading from cover crops. In general, on a daily basis, soils with cover crops are warmer at night and cooler at daytime compared with those without cover crops. On a seasonal basis, soils with cover crops are warmer in winter but cooler in spring, summer, and fall. Cover crops can change soil temperature by an average of $\pm 2\,°C$ (Blanco-Canqui & Ruis, 2020).

Furthermore, soil temperature in fields with growing cover crops or cover crop residue mulch does not fluctuate as much as that in fields without residue mulch. Cover crops can moderate soil temperature and thus reduce the maximum and minimum soil temperature, and so decrease the abrupt fluctuations in soil temperature. The increased temperature at night and decreased temperature at daytime have numerous agronomic and environmental implications. They can reduce frequency of natural cycles (e.g., freeze–thaw, wet–dry, hot–cool), evaporation, organic matter decomposition, and favor other dynamics processes. For example, soils under cover crop residue cover in summer can have greater water content due to cooler temperature and lower evaporation rates compared with soils without cover crop residues. A study in the eastern portion of the Canadian prairies found berseem clover (*Trifolium alexandrium* L.) cover crop increased soil temperature in late fall by $3\,°C$, which delayed soil freezing and reduced the depth of the frozen soil layer from 0.6 to 0.4 m in winter (Kahimba et al., 2008). The same study found that in spring, soils under cover crops warmed and thawed earlier than soils without cover crops, increasing water infiltration and improving drainage. In sum, research indicates that introducing cover crops can be a strategy to reduce heat extremes near the soil surface and develop microclimates and energy balances in croplands.

3.2.2 Soil Chemical Properties

A potential improvement in soil chemical properties following cover crop introduction can have significant implications for the management of soil health, soil fertility, and environmental quality. However, what do we really know about cover crop effects on the soil chemical health? Research indicates that cover crops generally improve soil chemical properties, particularly near the soil surface (Abdollahi and Munkholm, 2014; Blanco-Canqui and Jasa, 2019; Nunes et al., 2018; Santos et al., 2022). One of the key chemical properties is soil organic matter because changes in its concentration can directly or indirectly affect other soil properties. Cover crops generally increase soil organic matter concentration near the soil surface (<10 cm; Cates et al., 2019; Jilling et al., 2020). However, changes in organic matter concentration after cover crop adoption can be slow and often detectable only several years (>3 years) after cover crop adoption.

Cover crops can also influence the labile fractions of organic matter (Cates et al., 2019; Jilling et al., 2020). Some of these fractions include particulate organic matter (POM), free light fraction, and mineralizable C or easily oxidizable C (Sequeira et al., 2011). A study on a tropical soil found the addition of cover crops enhanced the potential of no-till soils to increase POM concentration in the upper 5 cm depth compared with no-till without cover crops (Zotarelli et al., 2007). Similarly, in temperate soils, addition of cover crops generally increases POM concentration (Cates et al., 2019; Jilling et al., 2020). Changes in particulate organic matter or labile C can be sensitive indicators of changes in soil health after cover crop introduction in the short term.

Cover crops can have rapid effects on soil nitrate and P concentrations. Growing cover crops utilize nitrate and available P and thus reduce nutrient concentration in the soil. The reduction in soil nitrate concentration has positive implications for reducing the potential of nitrate leaching (Abdalla et al., 2019; Blanco-Canqui, 2018). Cover crops retain soil N and P in their shoots when growing, but, after cover crop termination, these nutrients are returned to the soil via mineralization for the subsequent crops. Cover crop impacts on other chemical properties such as soil pH, cation exchange capacity (CEC), and electrical conductivity appear to be variable. Cation exchange capacity refers to the maximum quantity of exchangeable cations that a soil can retain. In tropical soils, several studies reported no effects of cover crops on CEC (Ensinas et al., 2016; Souza et al., 2021). Similarly, in temperate soils, cover crop effects on CEC can be mixed or minimal (Eckert, 1991; Sharma et al., 2018).

The mixed effects on CEC suggest that cover crops may not always increase CEC (Nunes et al., 2018; Santos et al., 2022). While the direct impact of cover crops on CEC can be small, the increased soil organic matter concentration with cover crops can positively impact CEC, soil pH, nutrient availability, and other chemical properties in the long term. Cover crops can at least maintain soil chemical properties by reducing losses of nutrients (Abdollahi and Munkholm, 2014). In general, cover crops can modify soil chemical properties.

3.2.3 Soil Biological Properties

Abundance, biomass, and diversity of soil organisms affect the health of soils. Available data indicate that cover crops generally improve soil biological processes and properties, and thus soil biological health (Kim et al., 2020). Cover crops could alter soil biological properties more rapidly than other soil properties.

Changes in soil biological properties are perhaps the most relevant indicators of changes in soil health as living systems. Soil organisms and their interactions influence organic matter decomposition, C and N cycling, nutrient provision, water cycling, C translocation and sequestration, fluxes of water, air, and gases in the soil, and many other soil processes. Thus, biological properties directly influence the delivery of soil ecosystem services. Soil organisms have more positive effects on soil ecosystem services than negative effects (Kladivko, 2001).

Despite their importance, soil measurements after cover crop introduction have often focused more on soil chemical properties and less on soil biological properties. This is because interest in soil characterization has been mainly driven by crop production goals. Soil samples are often analyzed for changes in chemical properties such as pH, CEC, EC, organic matter, macronutrients, and micronutrients. The limited or lack of data on soil biological properties can be, in part, attributed to the complexity of characterization, high cost of analysis, and difficulties with the interpretation of biological properties. Thus, a complete understanding of cover crop effects on soil biological properties is still evolving.

3.2.3.1 Microorganisms

A meta-analysis of global data found cover crops can increase microbial abundance, microbial activity, and microbial diversity compared to no cover crops (Kim et al., 2020; Table 3.2). Cover crops can have larger impacts on microbial abundance and activity than on microbial diversity. Another meta-analysis found cover crops can increase microbial biomass C and N relative to no cover crops (Muhammad et al., 2012; Table 3.2). An increase in soil microbial biomass C with cover crops indicates a potential increase in total soil C.

Table 3.2 Results from Two Meta-Analyses of Studies on Cover Crop Impacts on Soil Microbial Properties

Soil microbial parameter	Percentage of increase after cover crop introduction
Microbial biomass	27
Microbial activity	22
Microbial diversity	2.5
Microbial biomass C	39
Microbial biomass N	51
Microbial biomass C to N ratio	20

Notes: Muhammad et al. (2012); Kim et al. (2020).

Cover crops can also increase microbial activity parameters such as enzymes and enzyme activities, which directly influence decomposition of soil organic matter and cycling of nutrients. As an example, in the northern U.S. Great Plains, addition of cover crops increased enzyme activity by 1.1- to 1.4-fold across various sites after the first and second rotations in winter wheat–fallow rotations (Housman et al., 2021). Similarly, in the southern U.S. High Plains, cover crops increased microbial community size, fungal abundance, and enzyme activities after two and three years of cover crop introduction (Thapa et al., 2021). Cover crops can be particularly important in improving soil biological health in water-limited regions with low organic matter soils and degraded soils. Cover crops can increase microbial abundance by adding labile organic matter and increasing belowground biomass input.

3.2.3.2 Macroorganisms

Cover crops may not only improve biomass, activity, and diversity of soil microorganisms (Kim et al., 2020) but also those of macroorganisms including earthworms, termites, beetles, ants, and others (Kelly et al., 2021). However, cover crop studies have often focused more on microorganisms or microbial properties than on macroorganisms. As an example of positive effects on cover crops on soil macroorganisms such as earthworms, a summary of recent studies indicates cover crops can increase earthworm population regardless of earthworm species in most cases (Table 3.3). Earthworm abundance in fields with cover crops can be two- to six-fold greater than in field without cover crops (Table 3.3). Cover crops improve earthworm population by providing food resource and shelter while creating favorable microclimate for earthworm proliferation relative to fields without cover crops.

An increase in the population of large soil organisms with cover crops can exert major changes in soil health through bio-tilling, burrowing, mixing, ingesting crop residues, and decomposing soil organic matter. For instance, deep-burrowing earthworms (e.g., *Lumbricus terrestris* L.) can engineer deep burrows to translocate crop residues and thus soil C to deeper layers in the soil profile. Earthworms are responsible for the translocation of soil C to deeper layers knowing most C is concentrated near the soil surface (<10 cm depth), especially in no-till systems (Lorenz & Lal, 2005). Deep burrows also contribute to soil aeration, improved soil drainage, root growth, and other processes.

An increase in the number of earthworms with cover crops is strongly and positively correlated with soil C accumulation (Fonte et al., 2009) and water infiltration (Blanco-Canqui et al., 2011). Cover crop-induced increase in soil organic C and organic matter can improve soil aggregate stability and create favorable soil conditions for earthworm habitat. While burrows can favor rapid transport of water and groundwater, they may cause nutrient losses

Table 3.3 Cover Crop (CC) Impacts on Earthworm Abundance

Location	Tillage	Cover crop	Earthworm species	Abundance (No. m^{-2})
[a]California, U.S.	No-till	No CC	*Lumbricus terrestris*	48b
		Triticale, rye, common vetch, radish, and clover		98a
	Tilled	No CC		20b
		Triticale, rye, common vetch, radish, and clover		90a
[b]Ardabil, Iran	Tilled	No CC	Not available	5b
		Buckwheat		21ab
		Barley		31a
		Chickling pea		25a
		Buckwheat + barley		30a
		Buckwheat + chickling pea		23ab
		Barley + chickling pea		18ab
		Buckwheat + barley + chickling pea		27a
[c]Pampas region of Argentina	No-till	No CC	*Aporrectodea caliginosa* and *Octolasion cyaneum*	40b
		Grass CCs		45b
		Legume CCs		90a
[d]Dundee, UK	Tilled	Various species	Not available	No change
[e]Vienna, Austria	Tilled	Radish (low and high seeding rate), and black oat, and Sudan grass	*Aporrectodea caliginosa* and *Aporrectodea rosea*	Mixed

Means followed by different lowercase letters within each study are significantly different.
[a] Kelly et al. (2021).
[b] Ghahremani et al. (2021).
[c] Rodriguez et al. (2020).
[d] Holland et al. (2021).
[e] Euteneuer et al. (2020).

(e.g., leaching) to groundwater via preferential flow (Bertrand et al., 2015). At the same time, cover crops scavenge and retain nutrients, which can reduce the amount of nutrients available for leaching through earthworm burrows. This suggests that systems with cover crops can reduce nitrate leaching through earthworm burrows unlike systems with abundant earthworms but no cover crops.

3.3 Interconnectedness of Soil Health Parameters

While soil health indicators are often divided into groups such as physical, chemical, and biological properties, these properties are, in reality, all interconnected within the soil matrix. Soil properties are not independent but closely interrelated. Further interactions can occur in response to management shift. While interactions among soil properties can occur under any soil management practice, such interactions can be accentuated under cover crops as this practice increases both aboveground and belowground biomass input and thus organic matter, which could readily change some dynamic soil properties (e.g., wet aggregate stability, labile organic matter concentration, soil temperature, and others).

Some examples of positive feedback and interrelationships among properties include the following. One, an increase in cover crop belowground biomass amount can improve biological properties, which in turn can increase aggregate stability and thus macroporosity, leading to increased aeration or water infiltration. Two, cover crop-induced increase in organic matter can favor soil aggregation, while, in turn, stable aggregates can protect C and promote long-term C sequestration. A long-term study of summer cover crops under no-till winter wheat–grain sorghum rotation conducted in Kansas further illustrates these positive feedbacks. An increase in soil organic matter concentration with cover crops was significantly correlated with reduced soil compactibility, and increased aggregate stability and water infiltration (Blanco-Canqui et al., 2011). The interrelationship among soil properties in response to cover crop introduction show that soil properties coalesce in the soil environment to determine the extent of improvement in soil health and the degree of soil service delivery.

3.4 Managing Soil Health

It is important to stress that the extent of changes in soil properties after cover crop adoption can be highly variable. As mentioned earlier, cover crops do not appear to rapidly improve bulk density, dry aggregate stability, hydraulic conductivity, water retention capacity, and some chemical properties (e.g., pH, EC, CEC). Thus, cover crop impacts on soil properties need to be monitored on a site-specific basis. The question is: Why are the impacts of cover crops on soil properties inconsistent? Cover crop impacts on soil properties may be inconsistent due to numerous factors including cover crop biomass production, cover crop species, and initial soil C or fertility, soil texture, among others (Steele et al., 2012; Muhammad et al., 2012). For example, initial soil conditions (e.g., compaction, C level, texture) can affect cover crop performance. Not all soils will respond the same to cover crop introduction due to differences in soil properties at the starting point.

to manage the desired soil processes and properties. Cover crop species that produce large amounts of biomass in a short period of time can be desirable to improve or maintain soil properties. This is particularly critical when the residue amount left by the previous crop is limited.

One may consider that planting a mix of diverse cover crop species could concomitantly improve soil properties more than single species. However, research data indicate that mixtures of cover crops may be no better than monocultures for improving soil properties. A global meta-analysis found that soil properties did not differ between mixtures of cover crops and monocultures in 88% of comparisons (Florence & McGuire, 2020). In some cases, cover crop mixes can underperform monocultures. Based on a global meta-analysis, soil microbial C and N were greater in monocultures than in mixtures of cover crops (Muhammad et al., 2012). Mixes of cover crops dominated by legumes could support soil biology better than mixes dominated by grasses due to differences in residue quality (e.g., C to N ratio) or nutritional value (Roarty et al., 2017). The limited effects of cover crop mixes on soil properties can be due to the similarities in cover crop biomass production between mixes and single species as discussed in the previous chapter. Data indicate that planting single species cover crops can be cost-effective for improving soil properties relative to costly cover crop mixes.

3.4.4 Tillage System

Plowing cover crops into the soil as green manure can reduce benefits of cover crop residues for maintaining soil properties relative to no-till systems where cover crop residues remain on the soil surface. Intensive tillage can rapidly alter the integrity of cover crop roots, accelerating their decomposition, and potentially reduce soil organic matter fractions. For example, concentration of labile C fractions can be readily altered by tillage as these fractions are prone to rapid losses under soil disturbance. Moreover, intensive tillage can harm soil macroorganisms such as earthworms (Kladivko, 2001). Thus, no-till management can be a better strategy for cover crops after termination relative to tilled systems (Olson et al., 2014).

Adding cover crops to no-till can enhance the potential of no-till systems to improve soil properties (Blanco-Canqui et al., 2011). Indeed, in some cases, no-till may be better than tilled systems only when accompanied by cover crops (Hobley et al., 2018). Effects of reduced till on soil properties can be between no-till and conventionally tilled systems. No-till can have some challenges, as discussed later (Carver et al., 2022), but it is generally the best system to improve or maintain soil properties, especially when combined with cover crops.

It is important to note green manuring with cover crops can improve soil properties relative to no cover crops (Lopez-Vicente et al., 2021). Thus, tilled systems with cover crops can be better than those without cover crops. Experimental data on cover crops and soil properties under different tillage systems within the same experiment are, however, few (Olson et al., 2014). In general, tillage system effects on soil properties can be in this order: No-till cover crops > No-till alone > Reduced till cover crops > Reduced till alone > Conventional till cover crops > Conventional till alone.

3.4.5 Initial Soil Condition

Cover crops can have limited positive impacts on soil properties in temperate or highly fertile soils relative to degraded or low fertility soils. In other words, degraded or low fertility soils may benefit from cover crops more than highly productive soils such as those in temperate regions. Soil properties interact with climate in their response to cover crops. Also, intensively plowed soils with low initial C can benefit more from cover crops relative to no-till with already improved soil properties. This suggests adding cover crops to previously plowed soils and then managing such soils under no-till with cover crops can be a potential option to restore the degraded properties. Plowed or degraded soils possess more room for improvement than no-till soils (Kim et al., 2020).

Also, addition of high rates of inorganic fertilizers may mask the positive effects of cover crops on soil ecosystem services (Blanco-Canqui et al., 2011). Cover crops could have more positive effects, such as microbial biomass and diversity, under low than high inorganic fertilization (Kim et al., 2020). To date, most data on cover crops and soil properties are, however, from highly productive or high-input systems. The latter systems may not need cover crops as much as the low fertility or degraded soils. Overall, cover crops could improve soil properties and thus soil ecosystem services more rapidly in erosion-prone or low organic matter soils than in fertile or high-input systems although data on cover crop impacts on soil properties from degraded or low fertility soils are limited.

3.5 Summary

Does cover cropping improve soil health? Research shows that cover crops improve most soil physical, chemical, and biological properties, which are indicators of soil health, especially near the soil surface and in the long term. Cover crops can reduce soil penetration resistance, increase wet aggregate stability, macroporosity, water infiltration, organic matter concentration, microbial properties, and earthworm population, and moderate temperature. For example, soils under cover crops can be warmer in winter but cooler in spring, summer, and fall. Micro- and macro-organisms could respond more rapidly to cover crop introduction relative to other soil properties. Cover crops, however, appear to have some inconsistent effects on soil bulk density, dry aggregate stability, hydraulic conductivity, water content at field capacity and permanent wilting point, and plant available water, pH, cation exchange capacity (CEC), and electrical conductivity.

The degree to which cover crops impact soil properties depends on cover crop biomass production, initial soil C, soil fertility level, climate, and other factors. One key factor for changes in soil properties is the amount of cover crop biomass produced. If biomass production is low ($< 1\,Mg\,ha^{-1}$), then changes in soil properties may be minimal. Approximately $5\,Mg\,ha^{-1}$ of cover crop biomass production may be the optimum level although this level has not been established for different soils and climates.

Cover crops could have more beneficial effects on soil properties in degraded or low C soils than in fertile and high C soils, but most available data are from cover crops established on highly productive systems. It is important to note all soil properties are interrelated. For instance, an increase in soil organic matter can enhance soil biological activity and soil aggregation, which, in turn, can increase macroporosity, water infiltration, water retention, and plant available water. In general, cover crops can be a strategy to maintain or improve or manage soil physical, chemical, and biological properties as indicators of soil health.

References

Abdalla, M., Hastings, A., Cheng, K., Yue, Q., Chadwick, D., Espenberg, M., Truu, J., Rees, R. M., & Smith, P. (2019). A critical review of the impacts of cover crops on nitrogen leaching, net greenhouse gas balance and crop productivity. *Global Change Biology, 25*, 2530–2543.

Abdollahi, L., & Munkholm, L. J. (2014). Tillage system and cover crop effects on soil quality: I. Chemical, mechanical, and biological properties. *Soil Science Society of America Journal, 78*, 262–270.

Abdollahi, L., Munkholm, L. J., & Garbout, A. (2014). Tillage system and cover crop effects on soil quality: II. Pore characteristics. *Soil Science Society of America Journal, 78*, 271–279.

Bagnall, D. K., Morgan, C. L. S., Cope, M., Bean, G. M., Cappellazzi, S., Greub, K., Liptzin, D., Norris, C. L., Rieke, E., Tracy, P., Aberle, E., Ashworth, A., Bañuelos Tavarez, O., Bary, A., Baumhardt, R. L., Borbón Gracia, A., Brainard, D., Brennan, J., Briones Reyes, D., Bruhjell, D., Carlyle, C., Crawford, J., Creech, C., Culman, S,. Deen, W., Dell, C., Derner, J., Ducey, T., Duiker, S. W., Dyck, M., Ellert, B., Entz, M., Espinosa Solorio, A., Fonte, S. J., Fonteyne, S., Fortuna, A.-M., Foster, J., Fultz, L., Gamble, A. V., Geddes, C., Griffin-LaHue, D., Grove, J., Hamilton, S. K., Hao, X., Hayden, Z. D., Howe, J., Ippolito, J., Johnson, G., Kautz, M., Kitchen, N.,

Kumar, S., Kurtz, K., Larney, F., Lewis, K., Liebman, M., Lopez Ramirez, A., Machado, S., Maharjan, B., Martinez Gamiño, M. A., May, W., McClaran, M., McDaniel, M., Millar, N. Mitchell, J. P., Moore, P. A. Moore, A., Mora Gutiérrez, M., Nelson, K. A., Omondi, E., Osborne, S., Alcalá, L. O., Owens, P., Pena-Yewtukhiw, E. M., Poffenbarger, H., Ponce Lira, B., Reeve, J., Reinbott, T., Reiter, M., Ritchey, E., Roozeboom, K. L., Rui, I., Sadeghpour, A., Sainju, U. M., Sanford, G., Schillinger, W., Schindelbeck, R. R. Schipanski, M., Schlegel, A., Scow, K., Sherrod, L., Sidhu, S., Solís Moya, E., St. Luce, M., Strock, J., Suyker, A., Sykes, V., Tao, H., Trujillo Campos, A., Van Eerd, L. L. Verhulst, N., Vyn, T. J., Wang, Y., Watts, D., Wright, D., Zhang, T., & Honeycutt, C. W. (2022). Carbon-sensitive pedotransfer functions for plant available water. *Soil Science Society of America Journal, 86*, 612–629.

Balkcom, K. S., Duzy, L. M., Kornecki, T. S., & Price, A. J. (2015). Timing of cover crop termination: Management considerations for the southeast. *Crop, Forage & Turfgrass Managemen, 1*, 1–7.

Bertrand, M., Barot, S., Blouin, M., Whalen, J., de Oliveira, T., & Roger-Estrade, J. (2015). Earthworm services for cropping systems. A review. *Agronomy for Sustainable Development, 35*, 553–567.

Blanco-Canqui, H. (2018). Cover crops and water quality. *Agronomy Journal, 110*, 1633–1647.

Blanco-Canqui, H., Ferguson, R. B., Jin, V. L., Schmer, M. R., Wienhold, B. J., & Tatarko, J. (2014). Can cover crop and manure maintain soil properties after stover removal from irrigated no-till corn? *Soil Science Society of America Journal, 78*, 1368–1377.

Blanco-Canqui, H., Holman, J. D., Schlegel, A. J., Tatarko, J., & Shaver, T. (2013). Replacing fallow with cover crops in a semiarid soil: Effects on soil properties. *Soil Science Society of America Journal, 77*, 1026–1034.

Blanco-Canqui, H., & Jasa, P. (2019). Do grass and legume cover crops improve soil properties in the long term? *Soil Science Society of America Journal, 83*, 1181–1187.

Blanco-Canqui, H., Mikha, M., Presley, D. R., & Claassen, M. M. (2011). Addition of cover crops enhances no-till potential for improving soil physical properties. *Soil Science Society of America Journal, 75*, 1471–1482.

Blanco-Canqui, H., & Ruis, S. (2020). Cover crops and soil physical properties. *Soil Science Society of America Journal*, 1527–1576.

Carver, R. E., Nelson, N. O., Roozeboom, K. L., Kluitenberg, G. J., Tomlinson, P. J., Kang, Q., & Abel, D. S. (2022). Cover crop and phosphorus fertilizer management impacts on surface water quality from a no-till corn-soybean rotation. *Journal of Environmental Management.* http://dx.doi.org/10.1016/j.jenvman.2021.113818

Cates, A. M., Ruark, M. D., Grandy, A. S., & Jackson, R. D. (2019). Small soil C cycle responses to three years of cover crops in maize cropping systems. *Agriculture, Ecosystems and Environment, 286*, 106649.

Chen, G., & Weil, R. R. (2011). Root growth and yield of maize as affected by soil compaction and cover crops. *Soil and Tillage Research, 117*, 17–27.

Colombi, T. L. C., Torres, A. W., & Keller, T. (2018). Feedbacks between soil penetration resistance, root architecture and water uptake limit water accessibility and crop growth - a vicious circle. *Science of the Total Environment, 626*, 1026–1035.

Díaz-Zorita, M., & Grosso, G. A. (2000). Effect of soil texture, organic carbon and water retention on the compactibility of soils from the Argentinean pampas. *Soil and Tillage Research, 54*, 121–126.

Doran, J. W., & Zeiss, M. R. (2000). Soil health and sustainability: Managing the biotic component of soil quality. *Applied Soil Ecology, 15*, 3–11.

Eckert, D. J. (1991). Chemical attributes of soils subjected to no-till cropping with rye cover crops. *Soil Science Society of America Journal, 55*, 405–409.

Ensinas, S. C., Serra, A. P., Marchetti, M. E., da Silva, E. F., Lourente, E. R. P., do Prado, E. A. F., Matos, F. A., Altomar, P. H., Martinez, M. A., & Potrich, D. C. (2016). Cover crops affect the soil chemical properties under no-till system. *Australian Journal of Crop Science, 10*, 1104–1111.

Euteneuer, P., Wagentrist, H., Steinkellner, S., Fuchs, M., Zaller, J. G., Piepho, H.-P., & Butt, K. R. (2020). Contrasting effects of cover crops on earthworms: Results from field monitoring and laboratory experiments on growth reproduction and food choice. *European Journal of Soil Biology, 100*, 103225.

Fageria, N. K., Baligar, V. C., & Bailey, B. A. (2005). Role of cover crops in improving soil and row crop productivity. *Communications in Soil Science and Plant Analysis, 36*, 2733–2757.

Finney, D. M., Buyer, J. S., & Kaye, J. P. (2017). Living cover crops have immediate impacts on soil microbial community structure and function. *Journal of Soil and Water Conservation, 72*, 361–373.

Florence, A. M., & McGuire, A. M. (2020). Do diverse cover crop mixtures perform better than monocultures? A systematic review. *Agronomy Journal, 112*, 3513–3534.

Fonte, S. J., Winsome, T., & Six, J. (2009). Earthworm populations in relation to soil organic matter dynamics and management in California tomato cropping systems. *Applied Soil Ecology, 41*, 206–214.

Ghahremani, S., Ebadi, A., Tobeh, A., Hashemi, M., Sedghi, M., Gholipoouri, A., & Barker, A. V. (2021). Short-term impact of monocultured and mixed cover crops on soil properties, weed suppression, and lettuce yield. *Communications in Soil Science and Plant Analysis, 52*, 406–415.

Hobley, E., Garcia-Franco, N., Hübner, R., & Wiesmeier, M. (2018). Reviewing our options: Managing water-limited soils for conservation and restoration. *Land Degradation and Development, 29*, 1041–1053.

Holland, J., Brown, J. L., MacKenzie, K., Neilson, R., Piras, S., & McKenzie, B. M. (2021). Over winter cover crops provide yield benefits for spring barley and maintain soil health in northern Europe. *European Journal of Agronomy, 130*, 126363.

Housman, M., Tallman, S., Jones, C., Miller, P., & Zabinski, C. (2021). Soil biological response to multi-species cover crops in the northern Great Plains. *Agriculture, Ecosystems and Environment, 313*, 107373.

Irmak, S., Mohammed, A. T., Sharma, V., & Djaman, K. (2018). Impacts of cover crops on soil physical properties: Field capacity, permanent wilting point, soil-water holding capacity, bulk density, hydraulic conductivity, and infiltration. *Transactions of the ASABE, 61*, 1307–1321.

Jilling, A., Kane, D., Williams, A., Yannardell, A. C., Davis, A., Jordan, N. R., Koide, R. T., Mortensen, D. A., Smith, R. G., Snapp, S. S., Spokas, K. A., & Grandy, A. S. (2020). Rapid and distinct responses of particulate and mineral-associated organic nitrogen to conservation tillage and cover crops. *Geoderma, 359*, 114001.

Kahimba, F. C., Ranjan, R. S., Froese, J., Entz, M., & Nason, R. (2008). Cover crop effects on infiltration, soil temperature, and soil moisture distribution in the Canadian prairies. *Applied Engineering in Agriculture, 24*, 321–333.

Kaspar, T. C., Radke, J. K., & Laflen, J. M. (2001). Small grain cover crops and wheel traffic effects on infiltration, runoff, and erosion. *Journal of Soil and Water Conservation, 56*, 160–164.

Kelly, C., Fonte, S. J., Shrestha, A., Daane, K. M., & Mitchell, J. P. (2021). Winter cover crops and no-till promote soil macrofauna communities in irrigated, Mediterranean cropland of California. *Applied Soil Ecology, 166*, 104068.

Kim, N., Zabaloy, M. C., Guan, K., & Villamil, M. B. (2020). Do cover crops benefit soil microbiome? A meta-analysis of current research. *Soil Biology and Biochemistry, 142*, 107701.

Kladivko, E. J. (2001). Tillage systems and soil ecology. *Soil and Tillage Research, 61*, 61–76.

Koehler-Cole, K., Elmore, R. W., Blanco-Canqui, H., Francis, C. A., Shapiro, C. A., Proctor, C. A., Ruis, S. J., Heeren, D. M., Irmak, S., & Ferguson, R. B. (2020). Cover crop productivity and subsequent soybean yield in the western corn belt. *Agronomy, 112*, 2649–2663.

Lehmann, J., Bossio, D. A., Kögel-Knabner, I., & Rillig, M. C. (2020). The concept and future prospects of soil health. *Nature Reviews Earth Environment*. http://dx.doi.org/10.1038/s43017-020-0080-8

Lopez-Vicente, M., Gómez, J. A., Guzmán, G., Calero, J., & García-Ruiz, R. (2021). The role of cover crops in the loss of protected and non-protected soil organic carbon fractions due to water erosion in a Mediterranean olive grove. *Soil Research, 213*, 105119.

Lorenz, K., & Lal, R. (2005). The depth distribution of soil organic carbon in relation to land use and management and the potential of carbon sequestration in subsoil horizons. In D. L. Sparks (Ed.), *Advances in agronomy, advances in agronomy* (Vol. 88, pp. 35–66). Elsevier Academic Press Inc.

Minasny, B., & McBratney, A. B. (2018). Limited effect of organic matter on soil available water capacity. *European Journal of Soil Science, 69*, 39–47.

Muhammad, I., Wang, J., Khan, A., Ahmad, S., Yang, L., Ali, I., Zeeshan, M., Ullah, S., Fahad, S., Ali, S., & Zhou, X. B. (2012). Impact of the mixture verses solo residue management and climatic conditions on soil microbial biomass carbon to nitrogen ratio: A systematic review. *Environmental Science and Pollution Research, 28*, 64241–64252.

Nouri, A., Lee, J., Yin, X., Tyler, D. D., & Saxton, A. M. (2019). Thirty-four years of no-tillage and cover crops improve soil quality and increase cotton yield in Alfisols, southeastern USA. *Geoderma, 337*, 998–1008.

Nunes, M. R., van Es, H. M., Shindelbeck, R., Ristow, A. J., & Ryan, M. (2018). No-till and cropping system diversification improve soil health and crop yield. *Geoderma, 328*, 30–43.

Olson, K., Ebelhar, S. A., & Lang, J. M. (2014). Long-term effects of cover crops on crop yields, soil organic stocks and sequestration. *Open Journal of Soil Science, 4*, 284–292.

Qi, J., Jensen, J. L., Christensen, B. T., & Munkholm, L. J. (2021). Soil structural stability following decades of straw incorporation and use of ryegrass cover crops. *Geoderma, 406*, 115463.

Roarty, S., Hackett, R. A., & Schmidt, O. (2017). Earthworm populations in twelve cover crop and weed management combinations. *Applied Soil Ecology, 114*, 142–151.

Rodriguez, M. P., Domínguez, A., Moreira Ferroni, M., Wall, L. G., & Bedano, J. C. (2020). The diversification and intensification of crop rotations under no-till promote earthworm abundance and biomass. *Agronomy, 10*, 919.

Santos, G. G., Rosetto, S. C., Barbosa, R. S., Melo, N. B., Soares de Moura, M. C., Santos, D. P., Flores, R. A., & Collier, L. S. (2022). Are chemical properties of the soil influenced by cover crops in the cerrado/caatinga ecotone? *Communications in Soil Science and Plant Analysis, 53*, 89–103.

Sequeira, C. H., Alley, M. M., & Jones, B. P. (2011). Evaluation of potentially labile soil organic carbon and nitrogen fractionation procedures. *Soil Biology and Biochemistry, 43*, 438–444.

Sharma, V., Irmak, S., & Padhi, J. (2018). Effects of cover crops on soil quality: Part II. Soil exchangeable bases (potassium, magnesium, sodium, and calcium), cation exchange capacity, and soil micronutrients (zinc, manganese, iron, copper, and boron). *Journal of Soil and Water Conservation, 73,* 652–668.

Singh, J., Singh, N., & Kumar, S. (2020). X-ray computed tomography–measured soil pore parameters as influenced by crop rotations and cover crops. *Soil Science Society of America Journal, 84,* 1267–1279.

Souza, M., Müller, V. J., Kurtz, C., Ventura, B. S., Lourenzi, C. R., Ribeiro Lazzari, C. J., Ferreira, G. W., Brunetto, G., Loss, A., & Comin, J. J. (2021). Soil chemical properties and yield of onion crops grown for eight years under no-tillage system with cover crops. *Soil and Tillage Research, 208,* 104897.

Steele, M. K., Coale, F. J., & Hill, R. L. (2012). Winter annual cover crop impacts on no-till soil physical properties and organic matter. *Soil Science Society of America Journal, 76,* 2164–2173.

Thapa, V. R., Ghimire, R., Acosta-Martínez, V., Marsalis, M. A., & Schipanski, M. E. (2021). Cover crop biomass and species composition affect soil microbial community structure and enzyme activities in semiarid cropping systems. *Applied Soil Ecology, 157,* 103735.

van Bruggen, A. H. C., Goss, E. M., Havelaar, A., van Diepeningen, A. D., Finckh, M. R., & Morris, J. G. (2019). One health - cycling of diverse microbial communities as a connecting force for soil, plant, animal, human and ecosystem health. *Science of the Total Environment, 664,* 927–937.

Zhang, Z., Yan, L., Wang, Y., Ruan, R., Xiong, P., & Peng, X. (2022). Bio-tillage improves soil physical properties and maize growth in a compacted vertisol by cover crops. *Soil Science Society of America Journal, 86,* 324–337.

Zotarelli, L., Alves, B. J. R., Urquiaga, S., Boddey, R. M., & Six, J. (2007). Impact of tillage and crop rotation on light fraction and intra-aggregate soil organic matter in two Oxisols. *Soil and Tillage Research, 95,* 196–206.

4

Water Erosion

4.1 Overview

Increased extreme weather events with frequent and intense rainstorms coupled with anthropogenic activities (e.g., intensive tillage, limited crop residue cover) have heightened concerns over water erosion in sloping agricultural lands on a global scale (Li & Fang, 2016). For instance, a recent regional assessment using remote-sensing approaches showed that approximately 35% of sloping soils in the midwestern U.S. have lost their A-horizon due to combined effects of water and tillage erosion (Thaler et al., 2021). While the total amount of rainfall could remain the same or even decrease in some locations under fluctuating climate, the intensity of rainfall can be high, which can directly exacerbate water erosion risks. Water erosion rates can exceed 20 Mg ha^{-1} yr^{-1} in sloping and marginal croplands (Borrelli et al., 2017). For example, in some countries in the African continent, soil erosion can be as high as 300 Mg ha^{-1} yr^{-1} (Karamage et al., 2016). Thus, soil loss by water can readily exceed the average tolerable level of soil erosion (11 Mg ha^{-1} yr^{-1}) in erosion-prone, low fertility, and sloping agricultural lands.

The consequences of water erosion on soil productivity, environmental quality, and overall soil ecosystem services are well recognized. Water erosion reduces soil fertility, increases soil C loss, increases the risk of water pollution, reduces crop yields, and causes economic losses. For instance, because most of the soil C is concentrated in the upper 10 cm of the soil, erosion preferentially removes soil C and reduces C sequestration in croplands. Water erosion may increase soil C stocks at the downslope or depositional sites but may reduce soil C stocks at the eroding field sites where soil C is needed to maintain soil productivity, filter runoff, and improve soil properties. Most of all, increased nutrient runoff from agricultural lands is attributed to non-source pollution of downstream waters (Hanrahan et al., 2021).

Cover Crops and Soil Ecosystem Services, First Edition. Humberto Blanco.
© 2023 American Society of Agronomy, Inc. / Crop Science Society of America, Inc. / Soil Science Society of America, Inc. Published 2023 by John Wiley & Sons, Inc.

Figure 4.1 Winter rye cover crops can reduce water erosion in sloping croplands. Photo by H. Blanco.

Land use and management practices determine the rate of soil loss under intense rain events. Increasing soil cover and reducing soil disturbance can be critical to reduce or prevent water erosion. Adoption of no-till cropping systems can reduce both water and tillage erosion, but the rate of no-till adoption is still limited, depending on the county, state, and country (Blanco-Canqui and Ruis, 2020). Can cover crops be a solution to halt soil loss and thus improve declining soil ecosystem services? Since cover crops are used to literally cover soil, it is considered cover crops could be an effective management strategy to reduce water erosion in sloping fields (Figure 4.1). However, the extent to which cover crops can alter water quality parameters such as runoff volume, sediment loss, sediment-associated nutrients and nutrient runoff could vary with cover crop management and climate. Particularly, the impacts of cover crops on dissolved nutrients in runoff have not been much discussed. The sections below discuss how cover crops affect these water quality parameters.

4.2 Runoff

The rainfall rate of intense rainstorms can exceed the ability of the soil to absorb water. Intense rainstorms reduce the amount of rain that can infiltrate into the soil by causing rapid soil aggregate detachment and surface sealing compared

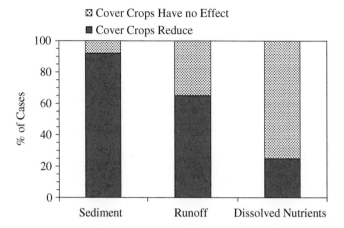

Figure 4.2 Cover crops reduce sediment loss and runoff in most cases but may reduce dissolved nutrients in runoff in only approximately 25% of cases according to a review by Blanco-Canqui (2018).

with low-intensity rainfalls. Excess rainfall increases risks of runoff and potential pollution of downstream waters. Increasing water infiltration rate with cover crops can be an option to reduce the amount of the rainwater available for runoff. The increased infiltration with cover crops can directly reduce runoff as long as the infiltration rate is greater or equal to the rainfall rate. Experimental data show cover crops can reduce runoff in approximately 65% of cases (Figure 4.2).

Cases studies in Table 4.1 show cover crops can reduce the amount of runoff by 8 to 93% when cover crops are growing. Because most studies have measured runoff only during the cover crop season, the extent to which cover crops could reduce runoff with time after termination is not well understood. How much of the cover crop benefits for reducing runoff carry over to the non-cover crop period is a remaining question. The reduction in runoff with cover crops can be due to the following mechanisms. First, cover crops intercept the erosive raindrops and dissipate their energy. Second, cover crops intercept runoff, promote ponding, and increase the opportunity time for water infiltration. This process delays runoff initiation and thus reduces the total amount of runoff (Table 4.1). Studies indicate cover crops can delay runoff start by 3 to 24 minutes relative to fields without cover crops (Table 4.2). Also, after runoff starts, runoff travels much slower over soils with cover crops than bare soils because cover crops increase surface roughness. While the magnitude of cover crop impacts on runoff reduction can be variable due to differences in cover crop management, research results indicate introduction of cover crops reduces runoff in most cases (Table 4.1).

Table 4.1 Cover Crop Impacts on Runoff and Sediment Loss Based on Recent Studies

Location	Crop	Cover Crop	Runoff (mm)	Sediment loss (Mg ha^{-1})
[a]Kansas, U.S.	Corn–soybean	No cover crop	Decreased or no change	0.5a
		Winter wheat, triticale, and rapeseed		0.2b
[b]Iowa, U.S.	Corn silage–soybean	No cover crop	27.3a	0.4a
		Winter rye	9.5b	0.1b
[c]Seville, Spain	Olive groves	No cover crop	9.5a	16.5a
		Various cover crop species	8.8b	9.2b
[d]Minas Gerais, Brazil	Olive groves	No cover crop	485a	304a
		Jack beans	150b	40b
		Millet	161b	64b
		Sunn hemp	33b	10b

Means with the same letter indicate no differences among cover crop treatments within the same study.
[a] Carver et al. (2022).
[b] Korucu et al. (2018).
[c] Lopez-Vicente et al. (2016, 2021).
[d] Beniaich et al. (2020).

Table 4.2 Cover Crop Impacts on Time to Runoff Initiation

Location	Crop	Cover crop	Time to runoff start (min)
[a]Valencia, Spain	Vineyards	No cover crop	3b
		Common vetch	12a
[b]Iowa, U.S.	Corn silage–soybean	No cover crop	5b
		Winter rye	11a
[c]Kansas, U.S.	Winter wheat–fallow	No cover crop	10b
		Winter lentil	20ab
		Spring triticale	14b
		Spring pea	13b
		Winter triticale	34a

Means with the same letter within the same study indicate no differences among cover crop treatments.
[a] Rodrigo-Comino et al. (2020).
[b] Korucu et al. (2018).
[c] Blanco-Canqui et al. (2013).

Third, cover crops add organic matter and promote aggregation and biological activity such as earthworm population and activity, which can improve macroporosity and thus infiltration. Fourth, cover crops anchor and hold crop residues from the previous crops on the soil surface that would otherwise be easily blown by wind or washed away by runoff. Thus, cover crops can be more effective than crop residues for erosion control. Crop residues may have reduced effectiveness for erosion control as rills could develop underneath the residues when compared with growing vegetation such as live cover crops (Kaspar et al., 2001). Overall, research data indicate cover crops can generally reduce the amount of runoff relative to croplands without cover crops by delaying runoff initiation, promoting infiltration, and improving related soil properties.

4.3 Sediment Loss

Runoff is the main agent that transports sediment. Thus, because cover crops generally reduce the total amount of runoff, cover crops can reduce sediment loss (Table 4.1). Growing cover crops coupled with residue mulch filter sediment and increase time for sediment deposition. Runoff leaving fields with cover crops is often less turbid than that leaving fields without cover crops. Cover crops can be more effective at reducing sediment loss than runoff. A global review found that cover crops can reduce sediment loss in over 90% of cases (Figure 4.2; Blanco-Canqui, 2018). The reduction in sediment loss ranges from 22 to 100% (Table 4.1; Blanco-Canqui, 2018). This indicates that cover crops can completely (100%) eliminate sediment loss or reduce it to negligible levels in some cases.

The same mechanisms that affect runoff can influence sediment loss. One of the leading soil processes by which cover crops can reduce sediment loss is soil aggregation. As discussed in Chapter 3, cover crops increase wet aggregate stability in most cases. The increased aggregate stability can increase the resistance of the soil to erosion. Increased aggregate stability is often attributed to increased soil organic C concentration. While cover crops may not alter total organic C in the short term, they could rapidly alter labile C fractions (e.g., particulate organic matter; Cates et al., 2019; Jilling et al., 2020). The labile C fractions can be more strongly correlated with soil aggregate stability than total C.

The no-till system is a proven practice to reduce water erosion. However, data show that adding cover crops to current no-till systems can enhance the potential of no-till to reduce runoff and sediment loss (Korucu et al., 2018; Carver et al., 2022). Thus, no-till combined with cover crops can be more effective at reducing water erosion than no-till alone. The addition of cover crops to tilled systems can also reduce water erosion when growing but the termination of cover crops via tillage can reduce or eliminate the benefits of cover crops for reducing water erosion. In general, cover crops can be an effective strategy to reduce losses of sediment by protecting the soil surface, improving soil aggregate stability, and reducing runoff, especially when combined with no-till technology.

4.4 Nutrient Loss

Runoff not only carries soil particles but also nutrients and pollutants associated with sediment and runoff water. While some nutrients and pollutants are transported with sediment as sediment-associated nutrients, a portion is carried in runoff as dissolved components. For example, runoff transports P as sediment-bound (e.g., particulate P) and as dissolved P. Losses of sediment-associated nutrients and dissolved nutrients depend on the amount of surface runoff and sediment loss. Sediment-associated nutrient loads (e.g., total P) can be correlated with the amount of eroded sediment (Liu et al., 2019; Carver et al., 2022). Because P is less mobile and more strongly bound to soil particles (e.g., clay) than total N, reducing sediment loss with cover crops can be particularly important to reduce P losses.

Some of the soluble nutrients in runoff include total P, NO_3-N, NH_4-N, and dissolved P. Experimental data indicate that cover crops can be less effective at reducing losses of dissolved nutrients than at reducing sediment-associated nutrients (Korucu et al., 2018). Cover crops can reduce transport of soluble nutrients in runoff in only approximately one-fourth of cases (Figure 4.2; Liu et al., 2019). The overall effectiveness of cover crops for reducing water erosion parameters can be in this order: Sediment losses > Runoff > Dissolved nutrients in runoff. Thus, the relatively sediment-free runoff water leaving fields with cover crops can still carry significant amounts of dissolved nutrients.

The reduction in soluble nutrient concentration in runoff in only approximately 25% and not in the rest (75%) of the cases indicates that the cover crop effect on nutrient runoff can be variable and highly site-specific. Research data also indicate cover crops can be more effective at reducing losses of NO_3-N and total N in runoff than losses of dissolved reactive P and total P concentrations (Blanco-Canqui, 2018; Liu et al., 2019). Indeed, cover crops could increase nutrient concentration in runoff when nutrients are released from their tissues as their residue is mineralized. This is discussed more in a later chapter.

4.5 Soil Carbon Loss

Water erosion has long been recognized as a major agent of soil degradation. However, most studies including those on cover crops report data only on runoff, sediment loss, and essential nutrients (N, P, and K), and not soil C loss

(Blanco-Canqui et al., 2013; Lopez-Vicente et al., 2021). Thus, soil C loss associated with sediment and runoff from croplands is often overlooked. Yet, the amount of soil C lost with erosion is an increasing concern as one of the pathways for C loss from terrestrial systems.

Cover crops can be strategy to reduce losses of C by reducing the amount of runoff and sediment loss. As an example, cover crops can reduce soil C loss with sediment by approximately 6.6 kg C ha^{-1} for the corn–soybean systems in Table 4.2 if we assume that soil lost contained 2% C (Carver et al., 2022; Korucu et al., 2018). Similarly, cover crops can reduce soil C loss by as much as 2.66 Mg C ha^{-1} for the orchard systems in Table 4.2 if we assume that soil lost contained 1% C (Beniaich et al., 2020; Lopez-Vicente et al., 2021). These estimates show cover crops can be effective at reducing C loss by reducing sediment loss relative to fields without cover crops.

Keeping soil C in place is critical to sequester C and reduce C emissions. Soil aggregate breakdown and transport during erosion accelerates oxidation of organic matter and releases C to the atmosphere. On average, approximately 20% of the eroded C is emitted to the atmosphere (Lal, 2003). Eroded C is an essential component of the global C budget. Soil C lost with eroded sediment is a net loss from croplands. Soil C is strongly correlated with other soil properties and it is thus essential to maintain and improve soil properties and thus soil health. Adding cover crops can be practice to halt the gradual depletion of soil C from croplands while reducing soil, water, and nutrient losses.

4.6 A Leading Factor of Water Erosion: Biomass Production

Cover crop biomass production is probably the single most important factor that determines the effectiveness of cover crops to reduce water erosion. Runoff and sediment loss exponentially decrease as cover crop biomass production increases (Rodrigo-Comino et al., 2020). Maintaining permanent vegetative cover in croplands by growing cover crops immediately after main crops can mimic natural systems with year-round cover. Cover crops provide additional biomass input and protect the soil during fallow periods when canopy cover from main crops is absent and crop residues are under decomposition.

Synchronizing the period of maximum surface cover from cover crops with the period when water erosion risks are the highest can be ideal to manage erosion. Soil is particularly susceptible to water erosion following the winter season due to limited crop residue cover and frequent freeze–thaw and wet–dry cycles that degrade soil properties during winter. Cover crop species (e.g., grasses) that overwinter and produce significant amounts of biomass in early spring can cover the soil and reduce soil erosion.

Both aboveground and belowground (roots) cover crop biomass production are critical for water erosion management. Cover crop roots can be as important as aboveground biomass production for mitigating erosion. Because roots are normally concentrated near the soil surface, their effects on stabilizing soil and reducing erosion can be large. Cover crop roots hold the soil in place and reduce soil detachment. Cover crops can also reduce sediment loss by stabilizing soil through their roots. Cover crops with fine roots can enmesh particles and hold soil together, reducing their erodibility. The effect of plant roots on stabilizing soil is well known (Gyssels et al., 2005). While the ability of cover crops to improve soil properties can be minimal or negligible in the short term (< 3 yr), the effect of cover crop roots on soil anchoring and erosion reduction can be immediate following successful cover crop establishment. The extent of cover crop root effects on erosion control depends on root characteristics including root diameter, amount, length, and density. Tap-rooted cover crops can drill and create large pore space, while fibrous roots can enmesh and hold the soil particles together (Chen & Weil, 2011).

Root channels can capture precipitation, increase water infiltration, and thus minimize surface runoff. Yu et al. (2016) compared the impact of 12 cover crop root systems on soil hydraulic conductivity and surface runoff and found cover crops distinctly influenced soil pore space, soil hydraulic conductivity, and runoff potential. Cover crops with coarse and high-density roots increased soil hydraulic conductivity and reduced runoff by 17% compared with cover crops with fine and low-density roots (Yu et al., 2016). Much focus has been on aboveground biomass characteristics and thus current understanding of different cover crop root traits and their implications for erosion mitigation, development of soil structure, and aggregate detachment are yet to be fully understood. Selection or development of cover crops with roots that penetrate deep into the soil can be important for increasing pore connectivity, drainage, hydraulic conductivity, and infiltration in the soil profile.

4.7 Cover Crops and Erosion-Prone Systems

The high effectiveness of cover crops for mitigating runoff and sediment loss indicates cover crops can be especially beneficial in sloping areas or erosion "hot spots" of the field where erosion control is needed (Panagos et al., 2021). Targeting water-erosion prone soils with cover crops in combination with no-till management can be a potential strategy for managing runoff and sediment losses in problem soils such as those in the midwestern U.S. (Thaler et al., 2021). While mechanical practices such as terraces can be measures to reduce soil loss, biological practices such as cover crops are better options, especially when cover crops have potential to provide multiple soil ecosystem services relative to mechanical structures.

The high effectiveness of cover crops for reducing runoff and sediment loss is not surprising. However, cover crops have not been widely adopted in water-erosion prone environments. While cover crops are highly effective at reducing soil loss (> 90%), the lack of direct economic return from cover crops often limits their widespread adoption although some moderate grazing of cover crops, as discussed later, is an evolving approach to achieve a dual purpose from cover crops: soil conservation as well as livestock production. Some of the systems that can particularly benefit from cover crop addition include low-biomass producing crops, corn silage and seed corn, when crop residue is removed, and orchard crops.

4.7.1 Low-Biomass Producing Cropping Systems

Low-biomass producing crops (e.g., soybean, cotton) can benefit from additional biomass input from cover crops. Research shows that runoff and soil losses can be greater after soybean than after corn in corn–soybean rotations, which are a common rotation in the midwestern U.S. Soybean not only produces less biomass but soybean residues decompose much faster than corn residues (Kaspar et al., 2001). Residues from legume cover crop (e.g., soybean) can contribute to increased soil aggregate stability but such positive effect is often short-lived when compared to soil aggregates formed under non-legume residues. Soil aggregate stability depends on both amount of residue produced and residue quality. A study in the midwestern U.S. reported that oat and winter rye cover crops when planted into soybean in late summer for three years reduced runoff and soil loss and that winter rye was more effective at reducing water erosion than oat due to its more prolonged growth and winter hardiness (Kaspar et al., 2001). In the southeastern U.S., Nouri et al. (2019) found that addition of hairy vetch and winter wheat cover crops compensated for the low cotton residue production and increased infiltration rate, cumulative infiltration, field-saturated hydraulic conductivity, and soil aggregate stability, reducing soil erodibility in a two-year study. Planting cover crops into or immediately after low-biomass producing crops can compensate for the low biomass input to reduce water erosion.

4.7.2 Corn Silage and Seed Corn

Cropping systems such as corn silage and seed corn that provide limited or no surface cover after harvest can use cover crops. These two systems leave the soil bare for a greater period than some of the other systems. Corn silage is a common forage crop in the U.S. especially for dairy farmers. However, corn silage harvest practically leaves the soil surface bare relative to corn grain harvest (Figure 4.3). Bare soils are highly susceptible to degradation of soil properties, surface runoff, soil loss, and nutrient runoff. Studies show that planting cover crops after corn

Figure 4.3 Oat cover crop planted in late summer and terminated in late fall by frost can provide abundant soil cover and reduce erosion risks the following spring (right) relative to no cover crops (left) in irrigated corn silage systems. (Blanco-Canqui et al., 2020; Photo by H. Blanco).

silage harvest can mitigate the negative effects of silage harvest. In the midwestern U.S., winter rye cover crops planted after corn silage harvest on fields with 8 to 15% slope reduced runoff by 67%, soil loss by 81%, P loss by 94%, and N loss by 83% under no-till continuous corn silage systems (Siller et al., 2016).

Adding cover crops to corn silage also increases soil organic matter, soil aggregate stability, water infiltration, and other soil properties. In some cases, cover crops may not improve corn silage production but can contribute to improved environmental quality and reduced fertilizer requirements. In a study in the western U.S. Corn Belt, corn silage harvest increased water erosion potential by reducing wet soil aggregate stability, organic matter concentration, and microbial biomass, but the same study reported that addition of an oat cover crop every other year after corn silage harvest did not overcome the adverse effects of corn silage on soil properties (Anderson et al., 2021). Based on this study, planting cover crop every year and using cover crops that overwinter can be a better strategy to potentially offset corn silage effects compared with planting every other year and using winter kill cover crops. Other studies found that planting winter rye after corn silage harvest offsets the negative impacts of corn silage in no-till continuous corn silage systems (Liesch et al., 2011).

4.7.3 Crop Residue Removal

Cover crops are a potential alternative to offset the erosion risks after crop residue removal. Crop residues are often baled or grazed to meet the demands for forage. Crop residue removal at high rates leaving less than approximately 4 Mg ha^{-1} of residues on the soil surface can increase the soil's susceptibility to water erosion. A study in the western U.S. Corn Belt reported that corn residue baling at rates above 50% increased runoff amount and sediment loss, which corroborates that soil loss is inversely correlated with surface residue cover (Kenney et al., 2015).

Cover crop biomass can partially replace the crop residue removed by baling and reduce erosion risks. Indeed, cover crops may not only replace the crop residue removed but also add extra belowground biomass (roots) to the cropping system, which can further stabilize and reduce soil erodibility. Cover crops can mitigate the potential adverse effects of crop residue removal on soil properties if cover crop biomass production is sufficient to supplant the amount of residue removed (Chahal & Van Eerd, 2020). Cover crops can be more effective at mitigating the negative effects of low or moderate rates (approximately 50%) of residue removal than high rates or residue removal.

4.7.4 Orchard Crops

Orchard systems including vineyards and citrus and olive groves which are common in Mediterranean environments, could benefit more from cover crops than cereal crops. Orchard and vineyard systems are often tilled between the rows of vines or trees, which can exacerbate erosion. Further, soils under orchards are often left bare and erosion rates are especially high during the first months and years during orchard establishment. Cerdà et al. (2021) reported extremely high (277 Mg ha^{-1} yr^{-1}) soil losses in citrus plantations in Mediterranean regions. The rate of soil loss from orchards can be several orders of magnitude greater than from most croplands under grain production.

Studies show that cover crops can lower the high erosion rates from orchard systems (Table 4.1). Soil loss from orchard systems can decrease by as much as 97% when cover crops are added (Beniaich et al., 2020). Well-established cover crops can be the top biological option for reducing soil losses in systems with limited surface cover and high erosion rates (Table 4.1). Similar to the impacts of cover crops on grain crops, addition of cover crops to orchard crops can be more effective at reducing soil loss than runoff.

When cover crops are introduced to orchard crops, soil structure, water transmission properties, and organic matter level improve (Cerdà et al., 2021; Rodrigo-Comino et al., 2020). Furthermore, a decrease in runoff with cover crops can contribute to water storage in regions with relatively limited rainfall input during the growing season. A review by Novara et al. (2021) discussed that cover crops can store water by reducing runoff in orchard crops. In regions with limited precipitation but intense seasonal rainstorms, growing cover crops for short periods of time for erosion control can be an alternative to reduce water use by cover crops for the main crops (Novara et al., 2021). It is important to note, however, cover crops alone may not completely eliminate or control soil erosion but can be a key companion practice to other conservation practices (e.g., no-till) to mitigate soil loss.

4.8 Summary

Cover crops are a useful management strategy to reduce runoff and soil losses under increasing extreme weather events with intense and localized rainstorms. Cover crops reduce runoff in 65% of cases and soil loss in 90% of cases. Runoff amount can decrease by 8 to 93% and soil loss by 22 to 100% with cover crops. Also, cover crops reduce sediment-associated nutrients but may have limited effects on reducing losses of dissolved nutrient in runoff. Cover crops can reduce dissolved nutrients in only 25% of cases. The order of effectiveness of cover crops for reducing water erosion parameters can be: Sediment loss > Runoff > Nutrient runoff. Cover crops can also reduce soil C losses associated with runoff and sediment.

Cover crops can intercept and dissipate the energy of erosive raindrops and delay runoff start. Also, soil aggregate stability, macroporosity, organic matter, hydraulic conductivity, infiltration, and other properties related to soil erodibility improve with cover crops. Moreover, cover crops can anchor and hold the soil in place through their roots although the critical role of cover crop roots in mitigating water erosion and improving soil properties is often overlooked.

Low-biomass producing crops (e.g., soybean, cotton), corn silage systems, fields after crop residue removal, and orchard crops (e.g., vineyards and citrus and olive groves) can especially benefit from cover crop use to reduce the high soil losses from these systems. Because cover crop biomass production determines the effectiveness of cover crops for erosion control, enhancing cover crop biomass production such as via early planting and late termination should be a priority.

References

Anderson, H. S., Johengen, T. H., Miller, R., & Godwin, C. M. (2021). Accelerated sediment phosphorus release in Lake Erie's central basin during seasonal anoxia. *Limnol. Oceanography.*, *66*, 3582–3595.

Beniaich, A., Silva, M. L. N., Guimarães, D. V., Bispo, D. F. A., Avanzi, J. C., Curi, N., Pio, R., & Dondeyne, S. (2020). Assessment of soil erosion in olive orchards (*Olea europaea* L.) under cover crops management systems in the tropical region of Brazil. *Revista Brasileira de Ciência do Solo, 44*, e0190088.

Blanco-Canqui, H. (2018). Cover crops and water quality. *Agronomy Journal, 110*, 1633–1647.

Blanco-Canqui, H., Drewnoski, M., Redfearn, D., Parsons, J., Lesoing, G., & Tyler, W. (2020). Does cover crop grazing damage soils and reduce crop yields? *Agrosystems Geosciences Environment, 3*, e20102.

Blanco-Canqui, H., Holman, J. D., Schelgel, A. J., Tatarko, J., & Shaver, T. M. (2013). Replacing fallow with cover crops in a semiarid soil: Effects on soil properties. *Soil Science Society of America Journal, 77*, 1026–1034.

Blanco-Canqui, H., & Ruis, S. (2020). Cover crops and soil physical properties. *Soil Science Society of America Journal*, 1527–1576.

Borrelli, P., Robinson, D. A., Fleischer, L. R., Lugato, E., Ballabio, C., Alewell, C., Meusburger, K., Modugno, S., Schütt, B., Ferro, V., Bagarello, V., Oost, K. V., Montanarella, L., & Panagos, P. (2017). An assessment of the global impact of 21st century land use change on soil erosion. *Nature Communications, 8*, 2013.

Carver, R. E., Nelson, N. O., Roozeboom, K. L., Kluitenberg, G. J., Tomlinson, P. J., Kang, Q., & Abel, D. S. (2022). Cover crop and phosphorus fertilizer management impacts on surface water quality from a no-till corn-soybean rotation. *Journal of Environmental Management.* https://doi.org/10.1016/j.jenvman.2021.113818

Cates, A. M., Ruark, M. D., Grandy, A. S., & Jackson, R. D. (2019). Small soil C cycle responses to three years of cover crops in maize cropping systems. *Agriculture, Ecosystems and Environment, 286*, 106649.

Cerdà, A., Novara, A., & Moradi, E. (2021). Long-term non-sustainable soil erosion rates and soil compaction in drip-irrigated citrus plantation in eastern Iberian Peninsula. *Science of the Total Environment, 787*, 147549.

Chahal, I., & Van Eerd, L. L. (2020). Cover crop and crop residue removal effects on temporal dynamics of soil carbon and nitrogen in a temperate, humid climate. *PLoS One, 15*. https://doi.org/10.1371/journal.pone.0235665

Chen, G., & Weil, R. R. (2011). Root growth and yield of maize as affected by soil compaction and cover crops. *Soil and Tillage Research, 117*, 17–27.

Gyssels, G., Poesen, J., Bochet, E., & Li, Y. (2005). Impact of plant roots on the resistance of soils to erosion by water: A review. *Progress in Physical Geography, 29*, 189–217.

Hanrahan, B. R., King, K. W., Duncan, E. W., & Shedekar, V. S. (2021). Cover crops differentially influenced nitrogen and phosphorus loss in tile drainage and surface runoff from agricultural fields in Ohio, USA. *Journal of Environmental Management, 293*, 112910.

Jilling, A., Kane, D., Williams, A., Yannardell, A. C., Davis, A., Jordan, N. R., Koide, R. T., Mortensen, D. A., Smith, R. G., Snapp, S. S., Spokas, K. A., & Grandy, A. S. (2020). Rapid and distinct responses of particulate and mineral-associated organic nitrogen to conservation tillage and cover crops. *Geoderma, 359*, 114001.

Karamage, J., Zhang, C., Ndayisaba, F., Shao, H., & Kayiranga, A. (2016). Extent of cropland and related soil erosion risk in Rwanda. *Sustainability, 8*, 609.

Kaspar, T. C., Radke, J. K., & Laflen, J. M. (2001). Small grain cover crops and wheel traffic effects on infiltration, runoff, and erosion. *Journal of Soil and Water Conservation, 56*, 160–164.

Kenney, I., Blanco-Canqui, H., Presley, D. R., Rice, C. W., Janssen, K., & Olson, B. (2015). Soil and crop response to Stover removal from rainfed and irrigated corn. *Global Change Biology. Bioenergy, 7*, 219–230.

Korucu, T., Shipitalo, M. J., & Kaspar, T. C. (2018). Rye cover crop increases earthworm populations and reduces losses of broadcast, fall-applied, fertilizers in surface runoff. *Soil and Tillage Research, 180*, 99–106.

Lal, R. (2003). Soil erosion and the global carbon budget. *Environment International, 29*, 437–450.

Li, Z. Y., & Fang, H. Y. (2016). Impacts of climate change on water erosion: A review. *Earth Science Reviews, 163*, 94–117.

Liesch, A. M., Krueger, E. S., & Ochsner, T. E. (2011). Soil structure and physical properties under rye-corn silage double-cropping systems. *Soil Science Society of America Journal, 75*, 1307–1314.

Liu, J., Macrae, M. L., Elliott, J. A., Baulch, H. M., Wilson, H. F., & Kleinman, P. J. A. (2019). Impacts of cover crops and crop residues on phosphorus losses in cold climates: A review. *Journal of Environmental Quality, 48*, 850–868.

Lopez-Vicente, M., García-Ruiz, R., Guzmán, G., Vicente-Vicente, J. L., Van Wesemael, B., & Gómez, J. A. (2016). Temporal stability and patterns of runoff and runon with different cover crops in an olive orchard (SW Andalusia, Spain). *Catena, 147*, 125–137.

Lopez-Vicente, M., Gómez, J. A., Guzmán, G., Calero, J., & García-Ruiz, R. (2021). The role of cover crops in the loss of protected and non-protected soil organic carbon fractions due to water erosion in a Mediterranean olive grove. *Soil Research, 213*, 105119.

Nouri, A., Lee, J., Yin, X., Tyler, D. D., & Saxton, A. M. (2019). Thirty-four years of no-tillage and cover crops improve soil quality and increase cotton yield in Alfisols, southeastern USA. *Geoderma, 337*, 998–1008.

Novara, A., Cerda, A., Barone, E., & Gristina, L. (2021). Cover crop management and water conservation in vineyard and olive orchards. *Soil and Tillage Research, 208*, 104896.

Panagos, P., Ballabio, C., Himics, M., Scarpa, S., Matthews, F., Bogonos, M., Poesen, J., & Borrelli, P. (2021). Projections of soil loss by water erosion in Europe by 2050. *Environmental Science & Policy, 124*, 380–392.

Rodrigo-Comino, J., Terol, E., Mora, G., Giménez-Morera, A., & Cerdà, A. (2020). Vicia sativa Roth. Can reduce soil and water losses in recently planted vineyards (*Vitis vinifera* L.). *Earth Systems Environment, 4*, 827–842.

Siller, A. R. S., Albrecht, K. A., & Jokela, W. E. (2016). Soil erosion and nutrient runoff in corn silage production with Kura clover living mulch and winter rye. *Agronomy Journal, 108*, 989–999.

Thaler, E. A., Larsen, I. J., & Yu, Q. (2021). The extent of soil loss across the US Corn Belt. *Proceedings of the National Academy of Sciences of the United States of America, 118*, 1–18.

Yu, Y., Loiskandl, W., Kaul, H. P., Himmelbauer, M., Wei, W., Chen, L., & Bodner, G. (2016). Estimation of runoff mitigation by morphologically different cover crop root systems. *Journal of Hydrology, 538*, 667–676.

5
Wind Erosion

5.1 Extent of Wind Erosion

Wind erosion is a growing global environmental concern, particularly in arid (<250 mm precipitation) and semi-arid (250–500 mm) environments (Vos et al., 2022). Increased erratic rainfall patterns and frequent droughts have increased concerns regarding wind erosion and dust emissions around the globe. Drier conditions not only increase soil erodibility but also limit biomass production for soil protection. Increased windstorms and dust loading reduce visibility for traffic and impact human health (Lambert et al., 2020).

Similar to water erosion, losses of soil C and nutrients associated with dust loading reduce soil C sequestration and long-term soil fertility and productivity (Darapuneni et al., 2021). Wind erosion is exacerbated by intensive tillage and limited crop residue input (Duniway et al., 2019; Sharratt et al., 2018). Further, agricultural expansion to marginal lands and grasslands is contributing to increased wind erosion risks in arid and semiarid environments (Lambert et al., 2020).

Estimates show that soil loss by wind ranges from 0.46 to 159 $Mg\,ha^{-1}\,yr^{-1}$ globally (Yang et al., 2021). The amount of soil lost by wind in arid and semiarid regions is greater than the rate of soil formation, which does not often exceed 1 $Mg\,ha^{-1}\,yr^{-1}$. Wind erosion rates vary across continents due to differences in agricultural practices and climatic conditions. They can be in this order: Africa > Asia > South America > Oceania > North America > Europe (Yang et al., 2021).

Conservation tillage such as no-till can reduce wind erosion, but even under no-till systems, wind erosion can be a concern if residue production or cover is limited especially during periods when main crops are absent. Indeed, soil erosion by wind from no-till soils can be as high as from tilled soils under limited residue cover. No-till technology can be better than conventional till to control wind

---

Cover Crops and Soil Ecosystem Services, First Edition. Humberto Blanco.
© 2023 American Society of Agronomy, Inc. / Crop Science Society of America, Inc.
/ Soil Science Society of America, Inc. Published 2023 by John Wiley & Sons, Inc.

Figure 5.1 Winter rye cover crop planted to manage wind erosion in the western U.S. Corn Belt. Photo by H. Blanco.

erosion only when there is sufficient residue cover in no-till soils. The question is: Can cover crops be used to significantly reduce wind erosion in arid and semi-arid environments where wind erosion predominates? Cover crops are a long-recognized practice to manage or control soil loss by wind (Figure 5.1). A recent survey of producers in the U.S. Midwest found that farmers value the use of cover crops to mitigate wind erosion, particularly in sandy soils (Yoder et al., 2021). However, the extent of cover crop impacts on wind erosion rates and soil erodibility by wind in water-limited regions is still unclear. This chapter discusses the magnitude of cover crop impacts on wind erosion based on experimental data.

5.2 Soil Loss

Studies measuring actual wind erosion rates under cover crops are relatively few. The available studies have quantified soil loss by wind using portable wind tunnels, Big Spring Number Eight (BNSE) samplers, and metal box samplers. Data from these studies indicate cover crops can generally reduce soil loss relative to no cover crops (Table 5.1). Cover crops can be particularly effective at reducing wind erosion in sandy soils, which are most susceptible to wind erosion, due to limited soil aggregation, low water holding capacity, and low organic C concentration. As an example, on a degraded sandy soil in southern Brazil, black oat and lupine cover

Table 5.1 Impact of Cover Crops on Wind Erosion

Location	Duration (yr)	Cover crop species	Cover crop condition	Impact of cover crop on wind erosion
[a]Rio Grande do Sul, Brazil	2	Black oat and lupine	Standing	Reduced by 93%
[b]New Mexico, U.S.	<1	Japanese millet, pearl millet, brown-top millet, and sorghum-sudangrass	Standing	Reduced by ~30% after 15 days of cover crop planting and by ~90% after 60–75 days of cover crop planting
[c]Texas, U.S.	<1	Sorghum-sudangrass and winter rye	Standing	Reduced by 2–3 orders of magnitude
[d]Washington, U.S.	2	Mustard	Green manure	Limited or no effect

[a] Rovedder and Eltz (2008).
[b] Darapuneni et al. (2021).
[c] Van Pelt and Zobeck (2004).
[d] Sharratt et al. (2018).

crops reduced the transport of soil particles with wind by 93% relative to no cover crops in a two-year study (Rovedder & Eltz, 2008). Also, on a sandy loam in the southern U.S. Great Plains, sorghum-sudangrass and barley cover crops reduced soil loss by wind by two- to three-fold compared with fields lacking cover crops (Van Pelt & Zobeck, 2004).

Cover crops can trap sediment by intercepting the blowing dust and reducing the flow of creeping and saltating particles. Cover crops can be particularly effective at reducing soil loss associated with creep and saltation relative to soil loss with suspension. Live cover crops can be more effective at intercepting wind and reducing dust generation than cover crop residues lying flat on the soil surface. Van Pelt and Zobeck (2004) observed that growing summer cover crops reduced soil loss by wind by two to three orders of magnitude, but frost-killed summer cover crops reduced it by one to two orders of magnitude. Also, cover crop rows oriented perpendicular to the dominant direction of the wind can be more effective at intercepting the high winds and buffering wind relative to cover crop rows oriented parallel to the wind direction. Similar to water erosion control, cover crop roots can contribute to wind erosion reduction by binding soil particles and anchoring the soil, thereby reducing the opportunity for dust generation. Cover crops can significantly increase the total amount of belowground biomass in cropping systems (Antosh et al., 2020).

The available data on soil loss by wind are from short-term (<2 year; Table 5.1) cover crop experiments. Multiple-year measurements under long-term cover crop experiments can help to better understand how cover crops perform under variable weather conditions. Soil surface conditions, cover crop growth, and biomass production often vary from year to year due to variable precipitation input. The effectiveness of cover crops to reduce wind erosion soil losses can also vary between standing cover crops and cover crop incorporated as green manure. Incorporating cover crops into the soil as green manure can reduce the effectiveness of cover crops to reduce wind erosion (Sharratt et al., 2018). While one may consider that green manure could reduce soil erodibility by increasing soil organic matter concentration and improving soil aggregation, data from some short-term cover crop studies suggest that such soil improvement may not overcome the adverse impacts of soil disturbance and accelerated crop residue decomposition that occur with green manuring. Overall, measured data on soil loss by wind indicate that standing cover crops when successfully established can protect soil from wind erosion.

5.3 Soil Erodibility

Some cover crop studies have measured soil properties that influence soil erodibility by wind instead of actual wind erosion amount. Such soil properties include dry aggregate stability and percent of erodible and non-erodible dry aggregates. The greater the proportion of dry stable aggregates and smaller the amount of erodible fraction, the greater the resistance of soil to wind erosion. The wind erodible fraction, which is the fraction most susceptible to wind erosion, refers to the proportion of soil dry aggregates with diameter less than 0.84 mm (Chepil, 1942).

Cover crop effects on soil dry aggregate stability and erodible fraction can be highly site-specific. A recent review of global data found cover crops have no effects on dry aggregate stability in 67% of cover crop versus no cover crop comparisons and that cover crops only increase dry aggregate stability in approximately one-fifth of cases (Blanco-Canqui et al., 2022). Similarly, cover crop effects on the wind erodible fraction can be minimal or mixed (Table 5.2). Further, the few available studies suggest cover crops could reduce the wind erodible fraction when combined with no-till but not when combined with tilled systems (Table 5.2). Soil disturbance, particularly intensive tillage, could erase the benefits of cover crops for soil aggregation.

Soil properties, especially physical properties (e.g., dry aggregate stability), are often slow to respond to cover crops. Cover crops could improve soil dry aggregate stability and other properties in the long-term (>5 year), but data from long-term cover crop experiments are few. A three-year study in the southwestern U.S. found

Table 5.2 Cover Crop Impacts on Soil Wind Erodible Fraction (<0.84 mm Dry Aggregates) as Affected by Different Cover Crops

Location	Soil	Duration (yr)	Cover crop species	Impact on wind erodible fraction
[a]Washington and Oregon, U.S.	Till	2	Mustard	No effect
[b]Saskatchewan, Canada	Till	33	Sweet clover	No effect
[c]Kansas, U.S.	No-till	5	Winter lentil, spring lentil, spring pea, winter triticale, and spring triticale	Reduced or no effect

[a] Sharratt et al. (2018).
[b] Campbell et al. (1993).
[c] Blanco-Canqui et al. (2013).

that mustard, cereal rye, and hairy vetch cover crops had no effect on soil dry aggregate stability in the first year after cover crop adoption but generally increased soil dry aggregate stability after the second year (Antosh et al., 2020). Also, a relatively long-term (six years) study in the semiarid Loess Plateau in China found that cover crops reduced the proportion of <0.25 mm dry aggregates and increased mean weight diameter of dry aggregates when cover crops were planted during the fallow period of winter wheat-fallow cropping systems (Zhang et al., 2019). The studies above indicate that cover crop use for multiple years is needed to realize the soil benefits from cover crops.

Experimental data suggest, in the short term (e.g., first year), cover crops could mainly mitigate wind erosion by physically covering the soil (Table 5.1) and not necessarily by improving soil properties (Table 5.2). As time increases after cover crop introduction, organic matter concentration and microbial activity often increase, which can contribute to soil aggregation. Labile C, fungal hyphae, mycorrhizal hyphae, and other soil biological components can chemically and biologically create bonds among soil particles to form stable aggregates. Indeed, Zhang et al. (2019) reported that a cover crop-induced increase in the proportion of dry macroaggregates (>5 mm diameter) was strongly and positively correlated with organic C concentration and enzyme activities after six years of cover crop management. In sum, cover crop effects on soil properties that influence soil erodibility by wind can be limited (Table 5.2), but cover crops can reduce wind erosion risks by covering the soil surface to protect soil from wind erosive forces (Table 5.1).

5.4 Managing Wind Erosion

Ensuring a uniform soil surface cover with cover crops via improved cover crop management can be key to reduce wind erosion in croplands with limited surface cover. However, cover crop performance depends on various factors including cover crop biomass production, management, and climate, which are discussed below. Interactions among the factors determine effects of cover crops on wind erosion relative to a single factor alone. For instance, larger amounts of cover crop biomass production can be needed to reduce wind erosion in wind erosion prone regions than in regions with limited wind erosion risks.

5.4.1 Biomass Production

Like water erosion, one of the key factors that affects cover crop effectiveness to reduce wind erosion is biomass production. If cover crops produce sufficient biomass to cover the soil during critical periods when soil is most susceptible to wind erosion, then cover crops can be an effective strategy to manage wind erosion. Studies from wind erosion-prone environments have found that cover crops can reduce soil loss by wind and improve dry aggregate stability and other soil properties when cover crops produce high amounts of biomass.

Approximately 4 Mg ha^{-1} of crop residues can be required to control wind erosion (Antosh et al., 2020; Darapuneni et al., 2021). In water-limited regions, some studies have used irrigation to establish cover crops (Darapuneni et al., 2021). If irrigated, cover crops can produce significant amounts of biomass. Indeed, the amount of cover crop biomass produced under irrigated conditions can be larger (3–20 Mg ha^{-1} of biomass; Antosh et al., 2020; Darapuneni et al., 2021) than that produced in high precipitation regions under rainfed conditions (Ruis et al., 2019).

The increased biomass production with irrigation can correspondingly result in improved soil properties. At two sites in a wind erosion-prone environment, irrigated mustard, rye, and vetch cover crops produced above 3 Mg ha^{-1} of biomass and increased mean weight diameter of dry aggregates by 1 mm (approximately 52%) at one site, while mustard increased it by 0.79 mm (33%) at the other site relative to no cover crops after three years (Antosh et al., 2020). These results suggest irrigated cover crops can be effective at improving dry soil aggregate stability if they produce significant amounts of biomass. A study in a rainfed environment found winter rye cover crop did not affect dry aggregate stability relative to no cover crop after three year when annual rye biomass production did not exceed 1 Mg ha^{-1} (Blanco-Canqui et al., 2014). This suggests that more than 1 Mg ha^{-1} of cover crop biomass production may be needed to significantly improve dry aggregate stability.

It is important to note, however, while irrigation can enhance cover crop biomass production, the declining water levels in aquifers in water-limited regions should be considered. Furthermore, irrigation can increase CO_2 emissions by pumping water and activating soil biological activities, which can have adverse effects on C sequestration and other ecosystem services from cover crops. Irrigating cover crops only a few times (deficit irrigation) may be an alternative to full irrigation to establish cover crops for wind erosion control in water-limited regions. In a two-year study in the arid southern U.S. Great Plains, Darapuneni et al. (2021) found that sorghum-sudangrass, pearl millet, Japanese millet, and brown-top millet cover crops produced large amounts of biomass under deficit irrigation (13.5 Mg ha^{-1} under deficit irrigation and 15.8 Mg ha^{-1} under full irrigation), which reduced soil loss by wind and improved soil dry aggregate stability. However, another study under rainfed conditions from the semiarid central U.S. Great Plains found that biomass production was modest for different cover crop species and thus they had limited or mixed potential to improve soil dry aggregate stability and reduce the wind erodible fraction (Blanco-Canqui et al., 2013). This comparison suggests that deficit irrigation could be an option to increase cover crop biomass production and reduce wind erosion in water-limited regions.

In some cases, a strategy to reduce soil water depletion by cover crops in water-limited regions would be to plant irrigated cover crops in strips similar to conservation buffer strips (Van Pelt & Zobeck, 2004). Well-designed strips at proper spacing could effectively intercept the wind, trap the blowing dust, and reduce wind erosion between the strips without covering the whole field with vegetation. Vegetative strips or barriers often protect soil downwind by approximately 10-fold their height. Thus, the spacing between strips of cover crops would have to be relatively narrow.

5.4.2 Cover Crop Species

Cover crop effects on soil loss by wind can also depend on cover crop species. Canopy cover, density, rate of growth, height, and thus biomass input vary among cover crop species. Rapid growth, winter hardiness, resistance to drought, and high biomass production are some of the desirable characteristics of potential species for wind erosion control. Cover crops that provide a dense surface cover after termination can reduce evaporation and conserve water while providing a uniform surface cover. Grass cover crops can be more effective than legume cover crops for reducing wind erosion. In the central U.S. Great Plains, a triticale cover crop was more effective than legume cover crops (winter lentil and Austrian pea) for reducing soil erodibility by wind due to its greater biomass production than other species (Blanco-Canqui et al., 2013).

Within the same species such as grasses, species that produce more biomass than others can be more effective at mitigating wind erosion. In the southern U.S. Great Plains, sorghum-sudangrass was more effective than other grass cover crops (Japanese millet, pearl millet, brown-top millet) for reducing dust transport as sorghum-sudangrass produced approximately twice as much biomass, provided better surface cover, had better density, and were taller than others (Darapuneni et al., 2021). The same study found the percentage of <0.25 mm dry aggregates under sorghum-sudangrass was approximately 20% lower than under Japanese millet, pearl millet, brown-top millet after two years of cover crop management. This suggests that sorghum-sudangrass cover crop can result in increased soil dry aggregate size and thus reduce wind erosion potential relative to other grass cover crops with lower biomass production.

Winter rye is another cover crop species commonly used to reduce wind and water erosion. A 12-year study from the Canadian Great Plains found addition of winter rye cover crop to low-biomass producing main crops such as potato and dry bean increased surface residue cover and reduced wind erosion risks (Larney et al., 2017). Based on the available studies, some of the potential grass cover crops that can be used for wind erosion control in wind erosion-prone regions include triticale, sorghum-sudangrass, and cereal rye (Tables 5.1 and 5.2).

5.4.3 Growth Stage and Seeding Rate

The stage of cover crop growth can affect cover crop effectiveness for wind erosion control as well as water erosion. Cover crop effectiveness for erosion reduction increases as the cover crop grows. In the southern U.S., cover crops under deficit irrigation reduced soil loss by 40 to 55% at 15 days after planting but reduced soil loss by 77 to 88% at 75 days after planting (Darapuneni et al., 2021). Cover crops near full maturity can be most effective at reducing wind erosion. Determining the right planting time of cover crops to ensure maximum soil coverage coinciding with the period when wind erosion risk is the highest would be ideal. In the U.S. Pacific Northwest where wind erosion is the highest between late summer and spring planting, Sharratt and Collins (2018) concluded that cover crops should be planted in late summer to match the wind erosion period.

Cover crop seeding rate is another factor that can influence cover crop effectiveness. Biomass production commonly increases with an increase in seeding rate up to an optimum level. In the U.S. southern Great Plains, cover crop ability to reduce wind erosion lasted longer when cover crops were seeded at 6 and 12 Mg ha^{-1} than when seeded at 3 Mg ha^{-1}, which was attributed to the greater biomass production under the higher seeding rate (Van Pelt & Zobeck, 2004). Low amounts of cover crop residue are prone to rapid decomposition, which can reduce its effectiveness to reduce wind erosion. Determining the optimum seeding rate for different cover crop species is important for producing sufficient biomass while reducing cover crop establishment costs.

5.4.4 Crop and Tillage Systems

Cropping systems with low biomass input (e.g., cotton, potato) can benefit the most from cover crops (Sharratt et al., 2018). Even high-biomass producing crops such as corn can benefit from cover crops for wind erosion reduction when corn residues are removed for off-farm uses (Figure 5.2). Additionally, cover crops should be planted every fallow period as the residual effects after cessation of cover crop use can quickly disappear. In the central U.S. Great Plains, cover crops planted during fallow in winter wheat-fallow systems tended to reduce wind erodible fraction when measured a few months after the cover crop termination but, one year after cover crop termination, the wind erodible fraction was similar between no cover crop and cover crop treatments (Blanco-Canqui et al., 2013). Also, similar to water erosion management, combining no-till with cover crops can be an effective strategy to halt

Figure 5.2 Soil surface exposed to water and wind erosion in early spring when corn silage was harvested the previous fall in a wind erosion prone soil in the western U.S. Corn Belt (Anderson et al., 2022). Photo by H. Blanco.

wind erosion relative to conventional till with cover crops due to limited or no soil disturbance under no-till.

5.4.5 Climate

Establishment and management of cover crops in wind erosion prone or water-limited environments can be challenging due to limited precipitation. Thus, the optimum amount of cover crop biomass production required for wind erosion control may not be achieved in arid and semiarid regions due to the limited water availability. Timely precipitation after cover crop planting is critical for the germination and establishment of cover crops. When precipitation amount is low or zero, cover crops may not germinate and thus fail to reduce or control wind erosion. This is a challenge in water-limited regions as wind erosion is often the highest during dry years when cover crops cannot be successfully established. As discussed earlier, deficit irrigation of cover crops can be a potential alternative to rainfed cover crops when timely precipitation is absent.

5.5 Summary

Experimental data on the magnitude of cover crop impacts on wind erosion for different climates and cover crop management scenarios are relatively few. Available data indicate that well-established cover crops can effectively cover the soil and reduce soil loss by wind. Data indicate, however, that cover crops do not always improve the soil properties (e.g., dry aggregate stability) to reduce soil erodibility by wind, particularly in the short term.

Cover crop effects on soil dry aggregate stability can vary from year to year within the same site due to differences in biomass input. Because cover crop biomass production is correlated with the potential of cover crops to reduce wind erosion and improve soil properties, growing cover crops in water-limited regions can be a challenge unless some irrigation is used. One or two irrigation events at cover crop establishment can be a viable option. Selection of cover crops with rapid growth, high density, and high biomass input is advisable for erosion control. Triticale, sorghum-sudangrass, and cereal rye are some of the high-performing grass cover crops used in the U.S. Great Plains.

Soils are particularly susceptible to wind erosion in early spring before main crops are established. Crop residue decomposition coupled with residue removal by wind during the winter months results in limited residue cover after winter. Achieving high cover crop biomass production during peak periods of wind erosion can be key to manage wind erosion in wind erosion prone environments.

References

Anderson, L., Blanco-Canqui, H., Drewnoski, M., & MacDonald, J. (2022). Cover crop grazing impacts on soil properties and crop yields under irrigated no-till corn-soybean. *Soil Science Society of America Journal, 86*, 118–133.

Antosh, E., Idowu, J., Schutte, B., & Lehnhoff, E. (2020). Winter cover crops effects on soil properties and sweet corn yield in semi-arid irrigated systems. *Agronomy Journal, 112*, 92–106.

Blanco-Canqui, H., Ferguson, R. B., Jin, V. L., Schmer, M. R., Wienhold, B. J., & Tatarko, J. (2014). Can cover crop and manure maintain soil properties after Stover removal from irrigated no-till corn? *Soil Science Society of America Journal, 78*, 1368–1377.

Blanco-Canqui, H., Holman, J. D., Schelgel, A. J., Tatarko, J., & Shaver, T. M. (2013). Replacing fallow with cover crops in a semiarid soil: Effects on soil properties. *Soil Science Society of America Journal, 77*, 1026–1034.

Blanco-Canqui, H., Ruis, S., Holman, H., Creech, C., & Obour, A. (2022). Can cover crops improve soil ecosystem services in water-limited environments? A review. *Soil Science Society of America Journal, 86*, 1–18.

Campbell, C. A., Moulin, A. P., Curtin, D., Lafond, G. P., & Townley-Smith, L. (1993). Soil aggregation as influenced by cultural practices in Saskatchewan: I Black Chernozemic soils. *Canadian Journal of Soil Science, 73*, 579–595.

Chepil, W. S. (1942). Measurement of wind erosiveness by dry sieving procedure. *Science in Agriculture, 23*, 154–160.

Darapuneni, M. K., Idowu, O. J., Sarihan, B., DuBois, D., Grover, K., Sanogo, S., Djaman, K., Lauriault, L., Omer, M., & Dodla, S. (2021). *Applied Engineering in Agriculture, 37*, 11–23.

Duniway, M. C., Pfenningwerth, A. A., Fick, S. E., Nauman, T. W., Belnap, J., & Barger, N. N. (2019). Wind erosion and dust from US drylands: A review of causes, consequences and solutions in a changing world. *Ecosphere, 10*, e02650.

Lambert, A., Hallar, A. G., Garcia, M., Strong, C., Andrews, E., & Hand, J. L. (2020). Dust impacts of rapid agricultural expansion on the Great Plains. *Geophysical Research Letters, 47*, e2020GL090347.

Larney, F. J., Pearson, D. C., Blackshaw, R. E., & Lupwayi, N. Z. (2017). Soil surface cover on irrigated rotations for potato (*Solanum tuberosum* L.), dry bean (*Phaseolus vulgaris* L.), sugar beet (*Beta vulgaris* L.), and soft white spring wheat (*Triticum aestivum* L.) in southern Alberta. *Journal of Soil and Water Conservation, 72*, 584–596.

Rovedder, A. P. M., & Eltz, F. L. F. (2008). Revegetation with cover crops for soils under arenization and wind erosion in Rio Grande do Sul state, Brazil. *Revista Brasileira de Ciência do Solo, 32*, 315–321.

Ruis, S. J., Blanco-Canqui, H., Creech, C. F., Koehler-Cole, K., Elmore, R. W., & Francis, C. A. (2019). Cover crop biomass production in temperate agroecozones. *Agronomy Journal, 111*, 1535–1551.

Sharratt, B. S., & Collins, H. P. (2018). Wind erosion potential influenced by tillage in an irrigated potato-sweet corn rotation in the Columbia Basin. *Agronomy Journal, 110*, 842–849.

Sharratt, B. S., McGuire, A., & Horneck, D. (2018). Early-season wind erosion influenced by soil-incorporated green manure in the Pacific northwest. *Soil Science Society of America Journal, 82*, 678–684.

Van Pelt, R. S. & Zobeck, T.M. (2004). *Effects of polyacrylamide, cover crops, and crop residue management on wind erosion.* ISCO – 13th International Soil Conservation Organization Conference Conserving Soil and Water for Society: Sharing Solutions. Brisbane, Australia.

Vos, H. C., Karst, I. G., Eckardt, F. D., Fister, W., & Kuhn, N. J. (2022). Influence of crop and land management on wind erosion from Sandy soils in dryland agriculture. *Agronomy, 12*, 457.

Yang, G., Sun, R., Jing, Y., Xiong, M., Li, J., & Chen, L. (2021). Global assessment of wind erosion based on a spatially distributed RWEQ model. *Progress in Physical Geography: Earth and Environment, 46*, 28–42.

Yoder, L., Houser, M., Bruce, A., Sullivan, A., & Farmer, J. (2021). Are climate risks encouraging cover crop adoption among farmers in the Southern Wabash River Basin? *Land Use Policy, 102*, 105268.

Zhang, D., Yao, Z., Chen, J., Yao, P., Zhao, N., He, W., Li, Y., Zhang, S., Zhai, B., & Wang, Z. (2019). Improving soil aggregation, aggregate-associated C and N, and enzyme activities by green manure crops in the loess plateau of China. *European Journal of Soil Science, 70*, 1267–1279.

6
Nutrient Losses

6.1 Implications of Nutrient Losses

Concerns over the rapid proliferation of toxic algae blooms that cause hypoxia and anoxia in fresh and marine water bodies are increasing around the globe. Decomposition of dense and harmful algae blooms results in high consumption of oxygen in aquatic ecosystems, leading to the development of hypoxic and anoxic zones. Depletion of oxygen to levels below $2\,\mathrm{mg\,L^{-1}}$ results in hypoxic zones, while the complete exhaustion of oxygen ($0\,\mathrm{mg\,L^{-1}}$) results in anoxic zones (Anderson et al., 2021; Hanrahan et al., 2021). These "dead zones" due to limited or no oxygen have adverse implications on ecosystem, human, and animal health. Some examples of hypoxic zones include the Baltic Sea, Gulf of Mexico, Chesapeake Bay, and Lake Erie (Hanrahan et al., 2021). Seasonal or recurring hypoxic zones often occur around the world, but the size and duration of hypoxic zones are increasing globally, particularly near intensively managed agricultural lands (Hanrahan et al., 2021).

The proliferation of algae blooms and formation of hypoxic and anoxic zones are primarily attributed to the excessive transport and accumulation of nutrients such as N and P from agricultural lands (Ali et al., 2021). Application of fertilizers and animal manure is important for maintaining or improving crop production, but their off-site transport in the form of N and P via leaching or runoff not only has on-site adverse impacts such as reduced soil fertility and productivity of croplands but also off-site impacts such as increased non-point source pollution of downstream lakes, rivers, streams, and coastal waters. Off-site movement of N and P in soluble forms from agricultural lands has long been recognized as a source of eutrophication in downstream water sources. However, extreme weather events such as increased heat waves and intense rainstorms have intensified the proliferation of algae blooms and water pollution in recent decades.

Cover Crops and Soil Ecosystem Services, First Edition. Humberto Blanco.
© 2023 American Society of Agronomy, Inc. / Crop Science Society of America, Inc. / Soil Science Society of America, Inc. Published 2023 by John Wiley & Sons, Inc.

First, increased water temperatures due to heat waves have contributed to the spread of algae blooms and hypoxic and anoxic zones to regions (e.g., northern latitudes) that were previously inhospitable to algae growth due to low temperatures. As an example, harmful algae blooms (*Alexandrium catenella* population) have been detected in the Alaskan Arctic where water temperature has increased by as much as 4 °C in the past two decades, making the environment conducive to algae growth (Anderson et al., 2021). Second, increased rainstorms are also contributing to increased nutrient leaching and nutrient runoff. The intensity of rainstorms has increased in recent decades, thereby resulting in heightened risks for leaching and runoff from agricultural watersheds (Ali et al., 2021; Hanrahan et al., 2021).

After decades of mitigation efforts, nutrient loadings into downstream waters persist leading to continued increases in algae blooms and hypoxic zones. This shows that past and current nutrient management scenarios have major shortcomings. This problem underscores the need for the redesign and establishment of more refined approaches for enhancing nutrient loss mitigation. The question is: Can cover crops reduce N and P loading from upstream agricultural watersheds into downstream waters? This inquiry is generating much interest to improve water quality and mitigate the growth of algae blooms and reduce the size and duration of hypoxic and anoxic zones. Cover crops can reduce N and P losses to aquatic systems if they reduce nutrient leaching, runoff, nutrient runoff, and losses of sediment and sediment-associated nutrients. As discussed in an earlier chapter, cover crops can be an effective practice to reduce sediment, sediment-associated nutrients, and runoff. This chapter discusses the effectiveness of cover crops on nutrient leaching reduction and transport of dissolved nutrients in runoff.

6.2 Nutrient Leaching

Because cover crops need nutrients for their growth, growing cover crops can capture a significant amount of surplus nutrients available after main crop harvest, and thus reduce the opportunity for nutrient leaching, particularly N. Depending on site-specific conditions, such as coarse-textured soils, leaching can be one of the main pathways for nitrate leaching. Global reviews of experimental data indicate cover crops can reduce nitrate leaching by an average of 54% in most cases (Blanco-Canqui, 2018; Thapa et al., 2018) in croplands.

However, the magnitude of cover crop impacts can vary, depending on cover crop biomass production, cover crop species, soil properties, precipitation amount, and others (Table 6.1). The potential of cover crops to reduce nitrate leaching

Table 6.1 Factors that can Affect Cover Crop Impacts on Nitrate Leaching

Factors	Nitrate Leaching
Cover crop biomass production	The greater the biomass production, the greater the reduction in nitrate leaching
Cover crop species	Non-legumes reduce nitrate leaching, but legumes have no effect
Cover crop planting date	Early planting (late summer or early fall) reduces nitrate leaching more than late planting
Cover crop termination date	Late termination (at main crop planting) can reduce nitrate leaching more than early termination
Tillage system	Tillage systems may not affect cover crop effects on nitrate leaching
Soil texture	Cover crops can reduce nitrate leaching more in coarse-textured than in fine-textured soils
Precipitation	Potential of cover crops to reduce nitrate leaching decreases as precipitation amount increases

Notes: Blanco-Canqui (2018); Thapa et al. (2018).

ranges from approximately 15 to 95%, which highlights the high variability in nitrate leaching reduction after cover crop introduction.

- Non-legume cover crops, particularly grass cover crops, can reduce nitrate leaching more than legume cover crops. Winter rye is a common cover crop that has high potential to scavenge N and reduce leaching. Well-established winter grass cover crops can be a strategy to reduce leaching in winter months or before crop planting when the potential for leaching is the highest. However, frost-kill legume and broadleaf cover crops could increase potential for nitrate leaching due to rapid residue decomposition and N release after termination.
- The potential for cover crops to reduce nitrate leaching increases with an increase in cover crop biomass production. A review by Thapa et al. (2018) found that the relationship between nitrate leaching and cover crop biomass production is not linear but quadratic. The reduction of nitrate leaching increased with biomass production but peaked between 4 and 8 Mg ha^{-1} of biomass production. This suggests that the optimum amount of biomass to capture substantial quantities of N can be between 4 and 8 Mg biomass ha^{-1}. Similar to other ecosystem services, if cover crops produce <1 Mg ha^{-1} of biomass, then the amount of N scavenged by these crops can be minimal (Blanco-Canqui, 2018). Cover crop biomass production often varies from year to year within the same site due to variable weather conditions, thus, so will the potential of cover crops to reduce nitrate leaching.

- Delaying cover crop termination until the main crop is planted can prolong the benefits of cover crops and reduce the opportunity for N losses. Nitrate leaching can be especially high during the period between early-terminated cover crops and main crop planting. Under typical management practices, cover crops are often terminated about a month in advance before crop planting, which leaves ample time for potential leaching of nitrates (Blanco-Canqui, 2018).
- Cover crop mixtures do not generally reduce nitrate leaching more than single-species cover crops because cover crop mixtures do not commonly produce more biomass than single-species cover crops (Florence & McGuire, 2020). High-biomass producing single species such as grass cover crops often reduce nitrate leaching more than cover crop mixtures. Among mixtures, non-legume cover crop mixtures can reduce nitrate leaching more than legume cover crop mixtures (Thapa et al., 2018).
- Cover crops can be more effective at reducing nitrate leaching in coarse-textured soils than in medium or fine-textured soils (Thapa et al., 2018). Cover crop effects not only vary with texture but also with other soil properties. An improvement in organic matter and soil aggregate stability with cover crops can enhance the ability to store nutrients and reduce leaching. Cover crops could be particularly beneficial to improve soil properties in degraded and sandy soils where soil properties need more improvement and leaching is of greater concern.
- An increase in precipitation amount can reduce cover crop effectiveness as there is more water volume to leach nutrients in the soil profile. Under low and moderate precipitation, cover crops can reduce nitrate leaching by using water that can be available for leaching.
- How tillage systems impact the potential of cover crops to reduce nitrate leaching is not conclusive (Blanco-Canqui, 2018; Thapa et al., 2018). This suggests that adding cover crops can be beneficial in both no-till and conventionally tilled systems. In some cases, no-till management may cause preferential or by-pass flow of water and nutrients when abundant large and vertical biopores are present in the root zone (Marhan et al., 2015).

Evidence supporting the high effectiveness of cover crops for mitigating nutrient leaching is ample. Thus, planting cover crops for using surplus N during periods when crops are not present can be ideal to reduce both nitrate losses and algal bloom risks. Winter grass cover crops that produce significant amounts of biomass (approximately $4\,Mg\,ha^{-1}$) and grow continuously between crop harvest and planting can be the top choice to minimize leaching of nutrients although achieving this amount of cover crop biomass production in cool temperate regions can be difficult. Reducing N leaching with cover crops not only mitigates algae blooms and water pollution but also retains N needed for the following crop, thereby potentially delivering multiple benefits including water quality improvement, soil fertility maintenance, and reduced farm production costs.

6.3 Dissolved Nutrients in Runoff

While, in general, cover crops are effective at reducing nitrate leaching, their ability to reduce the concentration of dissolved nutrients in runoff, especially dissolved reactive P, is questionable. This can be a concern because 35–40% of total P in runoff is present in the form of dissolved P. A four-year study in the central U.S. Great Plains under no-till corn–soybean rotation found that cover crops reduced sediment loss but increased dissolved reactive P losses and had inconsistent effects on total P losses (Carver et al., 2022). Also, a study comparing NO_3-N, total N, dissolved reactive P, and total P loads in surface runoff and subsurface (tile) drainage across 40 agricultural fields with and without cover crops in Ohio, U.S., found cover crops reduced tile NO_3-N and total N loads but increased tile dissolved reactive P loads and total P loads (Hanrahan et al., 2021). Additional data from Iowa and Kansas, U.S., further corroborate that cover crops reduce runoff and sediment loss but have inconsistent effects on the concentration of dissolved N and P in runoff relative to no cover crops (Table 6.2).

Cover crops may not only have limited ability to reduce transport of dissolved nutrients in runoff, but, in some cases, they can increase the concentration of dissolved nutrients such as dissolved P losses in runoff (White & Weil, 2011). Also, cover crop effects on nutrient losses can vary with the chemical species in that they can mitigate N losses more than P losses. Indeed, losses of P via surface

Table 6.2 Cover Crop Impacts on Dissolved Nutrients in Runoff

Location	Crop	Treatment	NH_4-N	NO_3-N	PO_4-P
				$mg\,L^{-1}$	
[a]Iowa, U.S.	Corn silage–soybean	No cover crop	9.90a	0.40	21.20a
		Winter rye	3.90b	0.20	9.30b
				$kg\,ha^{-1}$	
[b]Kansas, U.S.	Winter wheat–fallow	No cover crop	0.20	5.96a	0.30
		Winter lentil	0.09	3.36ab	0.17
		Spring triticale	0.19	2.55ab	0.22
		Spring pea	0.04	1.66b	0.06
		Winter triticale	0.12	1.41b	0.12

Means with different letters indicate differences among cover crop treatments within the same study.
[a] Korucu et al. (2018).
[b] Blanco-Canqui et al. (2013).

subsurface (tile) drainage are an increasing concern even in fields with cover crops (Hanrahan et al., 2021). Artificial drainage in fields with and without cover crops can facilitate a rapid connectivity of soil P with groundwater. Also, the preferential flow through earthworm burrows and root channels in no-till soils can increase losses of nutrients to tile drains (Macrae et al., 2021). Losses of P may not have large on-site adverse effects on crop production as P is required for plant growth in lower amounts than N, but they can have major off-site adverse effects by accelerating eutrophication and algal growth in downstream aquatic ecosystems. In sum, experimental data indicate cover cover crops can reduce nitrate leaching more than dissolved nutrients (e.g., dissolved P) in runoff.

6.4 Nutrient Release from Cover Crops

Cover crops could result in increased dissolved nutrients in rainfall and snowmelt runoff due to freeze–thaw cycles and fresh biomass decomposition (White & Weil, 2011). This is particularly true in cold climates. Frequent and abrupt freeze–thaw cycles can harm cover crop stems and leaves and cause the release of "water-extractable" P (Cober et al., 2018; Liu et al., 2019). Freezing temperatures cause the formation of ice crystals inside cover crop tissues, which damage the cells and release nutrients during thawing. Literature also indicates that the impact of freeze–thaw cycles can be greater on P concentration than on total N, NO_3, and NH_4 concentrations although the mechanisms for this difference are somewhat unclear. The magnitude of the impact of freeze–thaw cycles on nutrient release from cover crops can depend on climate, cover crop species, and cover crop termination practices, among others.

- Cold climates in northern latitudes can undergo frequent freeze–thaw cycles, which can cause larger release of dissolved nutrients than in regions with mild climates. For example, in a review, Liu et al. (2019) discussed that cover crops released more water-extractable P when cover crops were exposed to −16 °C than when exposed to −4 to −5 °C. Cover crops can release 40% or more of the total P during freeze–thaw cycles. In cool temperate regions, freeze–thaw cycles can be the main mechanism for the release of P from cover crops, while in mild and warm climates, rapid cover crop residue decomposition can be the main pathway for P release.
- The amount of water-extractable P released from cover crops not only depends on the frequency of freeze–thaw cycles but also on cover crop species. Winter-kill cover crop species can release more water-extractable P than winter hardy species. Also, water-extractable P release from brassica cover crops can be greater than from non-brassica cover crops as brassicas often have

higher P concentration. In southwestern Ontario, Canada, Cober et al. (2018) found oilseed radish and oat cover crops released more water-extractable P than cereal rye, red clover, and hairy vetch. This suggests that selection and development of cover crop species resistant to rapid nutrient release can be critical to reduce P release from cover crops and potential losses of P in runoff. It is important to note when freeze–thaw cycles and low temperatures are frequent, any cover crop species can release significant amounts of P.

- Live cover crops can be more resistant to resist freeze–thaw cycles than freshly terminated cover crops (Cober et al., 2018). Terminating cover crops at the onset of freeze–thaw cycles can greatly increase P release relative to cover crops terminated long before frost. Fresh plant material will release more P than dry crop residues due to its higher total P concentration. Also, cover crops terminated via freezing releases more P than those terminated via herbicides or clipping as freezing damages plant cells and accelerates release of nutrients (Liu et al., 2019).
- The larger the amount of cover crop biomass produced, the greater the amount of nutrient released to runoff. It is thus important to note that the idea of increasing cover crop biomass production for increasing ecosystem service benefits may not entirely apply for reducing the concentration of dissolved P in runoff in cold climates. However, the increased cover crop biomass production can also reduce the total amount of runoff, which can reduce nutrient runoff relative to field without cover crops.
- Losses of dissolved nutrients in runoff are positively correlated with precipitation amount and snowmelt runoff. Across 30 agricultural fields in Saskatchewan, Manitoba, and Ontario, Canada, losses of P in runoff varied with differences in climatic conditions and that snowmelt runoff was the main factor that determined P losses (Liu et al., 2021). In some cases, even if precipitation amount is low, losses of P can still be high if rains are intense.
- Differences in soil physical, chemical, and biological properties, land topography, cropping system, and climate among agricultural source areas determine the magnitude of nutrient losses and effectiveness of cover crops (Ni et al., 2020). For example, losses of dissolved P increase with increased soil test P and soil pH. Also, soils with limited soil aggregation and low macroporosity and water infiltration rates can be highly susceptible to water erosion and thus nutrients losses in runoff.

While the above points provide some insights into P release in runoff from cover crops, most studies, particularly those on the impact of freeze–thaw cycles on nutrient release, are from greenhouse experiments testing different levels of temperature, cover crops species, termination methods, and other factors. How cover crops interact with climate and termination methods under field conditions and contrasting environments in their impacts on nutrient release needs further assessment.

6.5 Management Implications

Losses of nutrients depend on the availability of nutrients near the soil surface and precipitation amount. Thus, lowering the amount of nutrients near the soil surface and reducing runoff to negligible levels during the fallow period with cover crop adoption can be strategies to reduce off-site nutrient transport. For example, the optimum level of available P for crop production can range from 30 to 50 mg kg^{-1} (Ni et al., 2020). Available P in excess of the amount above can be available for runoff. Adding cover crops can reduce runoff by increasing water infiltration and thus reducing the amount of precipitation water available for runoff.

Cover crops are often considered as the best strategy to mitigate nutrient losses while delivering other ecosystem services. However, the limited ability of cover crops to mitigate nutrient runoff in addition to the potential nutrient release from cover crops into runoff, especially in cold climates, suggests that management of cover crops and other field practices (e.g., fertilization) should be further refined (Figure 6.1). One of the companion practices that deserves emphasis is fertilizer management (4Rs) via right placement (e.g., broadcast, injection), right application time (e.g., fall, spring), right type (e.g., slow- versus fast-release), and right amount. Cover crops paired with the 4Rs can be a better option than either the use of cover crop alone or 4Rs alone. For example, deep or subsurface placement of fertilizers or animal manure can be a critical step under no-till systems (Smith et al., 2017). Timing of the subsurface injection of nutrients needs also consideration. Carver et al. (2022) found that losses of total P and dissolved reactive P in runoff were lower when P was injected in spring than in fall. Soils are normally wetter and more prone to water erosion in spring than in fall.

The variable and limited effect of cover crops on nutrient runoff further suggests that cover crops should be combined with other conservation practices to enhance the potential of cover crops to reduce nutrient losses and the development of hypoxic zones (Macrae et al., 2021). Among such conservation practices include no-till, reduced till, vegetative filter strips, grass hedges, riparian buffers, and grass waterways (Bosch et al., 2013). Also, the combination of cover crops with no-till and conservation buffers can be more effective at reducing nutrient loading when targeted to the critical source areas of nutrients than when established at "random" in agricultural lands where this combination may only reduce a small fraction of nutrient losses (Bosch et al., 2013). Selecting or identifying the most appropriate and cost-effective set of companion practices to cover crops should be the goal to achieve the pollution reduction targets with cover crops (Figure 6.1).

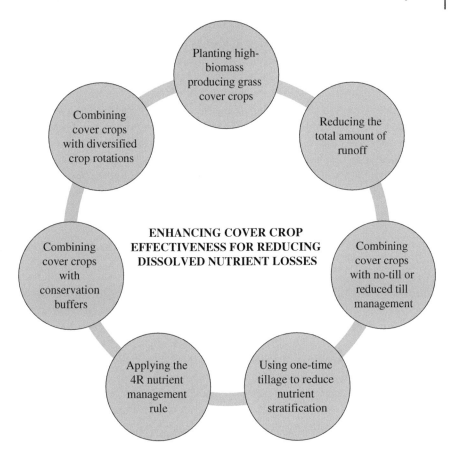

Figure 6.1 Some potential strategies for enhancing the potential of cover crops to reduce losses of dissolved nutrients.

6.6 Nutrient Stratification

The discussion above leads to the question: Is combining cover crops with no-till an option to enhance no-till potential to reduce dissolved nutrients in runoff? Data show that this combination can reduce water erosion and reduce nutrient leaching but may not always control losses of dissolved nutrients in runoff (Cober et al., 2018; Korucu et al., 2018; Liu et al., 2019). One challenge with no-till management is the vertical stratification of nutrients in the soil profile due to reduced or no soil mixing. As a result, nutrients in no-till systems are mostly concentrated near the soil surface (upper 10 cm soil). The enrichment of nutrients in the upper few centimeters of the soil surface in no-till systems often leads to high nutrient concentration in runoff.

Relatively immobile nutrients such as P are particularly prone to significant stratification under no-till management. The net effect of no-till management on dissolved nutrient losses even when combined with cover crops appears to be debatable. Current cover crop management strategies may need modification to enhance no-till potential to control losses of dissolved nutrients in runoff. Incorporation of deep-rooted cover crops into no-till systems could be an option to reduce nutrient stratification, promote nutrient transfer to deeper layers, and reduce nutrient losses in surface runoff. As discussed above, fertilizer management within the 4Rs nutrient stewardship accompanied by other conservation measures such as conservation buffers can be key to complement cover crop benefits under no-till management.

6.7 Summary

Non-point source pollution with N and P from agricultural lands increases risks of algae blooms and development of hypoxic and anoxic zones. Can cover crops address this growing environmental problem? Cover crops generally reduce nutrient leaching (approximately 55%) but may have limited potential for reducing concentration of dissolved nutrients in runoff. Indeed, concentration of dissolved nutrients in runoff from fields with cover crops could be higher than that from fields without cover crops. Cover crop residue decomposition coupled with freezing and thawing of cover crops can release nutrients (e.g., P) into runoff, particularly in cool climates with frequent freeze–thaw cycles. An increase in cover crop biomass production may not necessarily reduce losses of dissolved P.

However, note that while concentrations of dissolved nutrients in runoff with cover crops may be high, cover crops can reduce the amount of nutrients lost from a field by reducing the total amount of runoff. Winter hardy grass species such as cereal rye can be a potential species for reducing nutrient leaching and nutrient runoff in temperate regions. A cover crops accompanied by no-till, improved fertilizer management, and conservation buffers can be a strategy to significantly reduce nutrient loadings from agricultural lands when targeted to critical source areas of nutrients.

References

Ali, H., Fowler, H. J., Lenderink, G., Lewis, E., & Pritchard, D. (2021). Consistent large-scale response of hourly extreme precipitation to temperature variation over land. *Geophysical Research Letters, 48*, e2020GL090317.

Anderson, D. M., Fachon, E., Pickart, R. S., Lin, P., Fischer, A. D., Richlen, M. L., Uva, V., Brosnahan, M. L., McRaven, L., Bahr, F., Lefebvre, K., Grebmeier, J. M.,

Danielson, S. L., Lyu, Y., & Fukai, Y. (2021). Evidence for massive and recurrent toxic blooms of Alexandrium catenella in the Alaskan Arctic. *Proceedings of the National Academy of Sciences of the United States of America, 118*, e2107387118.

Blanco-Canqui, H. (2018). Cover crops and water quality. *Agronomy Journal, 110*, 1633–1647.

Blanco-Canqui, H., Holman, J. D., Schelgel, A. J., Tatarko, J., & Shaver, T. M. (2013). Replacing fallow with cover crops in a semiarid soil: Effects on soil properties. *Soil Science Society of America Journal, 77*, 1026–1034.

Bosch, N. S., Allan, J. D., Selegean, J. P., & Scavia, D. (2013). Scenario-testing of agricultural best management practices in Lake Erie watersheds. *Journal of Great Lakes Research, 39*, 429–436.

Carver, R. E., Nelson, N. O., Roozeboom, K. L., Kluitenberg, G. J., Tomlinson, P. J., Kang, Q., & Abel, D. S. (2022). Cover crop and phosphorus fertilizer management impacts on surface water quality from a no-till corn-soybean rotation. *Journal of Environmental Management, 301*, 113818.

Cober, J. R., Macrae, M. L., & Van Eerd, L. L. (2018). Nutrient release from living and terminated cover crops under variable freeze-thaw cycles. *Agronomy Journal, 110*, 1036–1045.

Florence, A. M., & McGuire, A. M. (2020). Do diverse cover crop mixtures perform better than monocultures? A systematic review. *Agronomy Journal, 112*, 3513–3534.

Hanrahan, B. R., King, K. W., Duncan, E. W., & Shedekar, V. S. (2021). Cover crops differentially influenced nitrogen and phosphorus loss in tile drainage and surface runoff from agricultural fields in Ohio, USA. *Journal of Environmental Management, 293*.

Korucu, T., Shipitalo, M. J., & Kaspar, T. C. (2018). Rye cover crop increases earthworm populations and reduces losses of broadcast, fall-applied, fertilizers in surface runoff. *Soil and Tillage Research, 180*, 99–106.

Liu, J., Elliott, J. A., Wilson, H. F., Macrae, M. L., Baulch, H. M., & Lobb, D. A. (2021). Phosphorus runoff from Canadian agricultural land: A cross-region synthesis of edge-of-field results. *Agricultural Water Management, 255*, 107030.

Liu, J., Macrae, M. L., Elliott, J. A., Baulch, H. M., Wilson, H. F., & Kleinman, P. J. A. (2019). Impacts of cover crops and crop residues on phosphorus losses in cold climates: A review. *Journal of Environmental Quality, 48*, 850–868.

Macrae, M., Jarvie, H., Brouwer, R., Gunn, G., Smith, D., Reid, K., Joosse, P., King, K., Kleinman, P., Williams, M., & Zwonitzer, M. (2021). One size does not fit all: Toward regional conservation practice guidance to reduce phosphorus loss risk in the Lake Erie watershed. *Journal of Environmental Quality, 50*, 1–18.

Marhan, S., Auber, J., & Poll, C. (2015). Additive effects of earthworms, nitrogen-rich litter and elevated soil temperature on N_2O emission and nitrate leaching from an arable soil. *Applied Soil Ecology, 86*, 55–61.

Ni, X., Yuan, Y., & Liu, W. (2020). Impact factors and mechanisms of dissolved reactive phosphorus (DRP) losses from agricultural fields: A review and synthesis study in the Lake Erie basin. *Science of the Total Environment, 714*, 136624.

Smith, D. R., Huang, C., & Haney, R. L. (2017). Phosphorus fertilization, soil stratification, and potential water quality impacts. *Journal of Soil and Water Conservation, 72*, 417–442.

Thapa, R., Mirsky, S. B., & Tully, K. L. (2018). Cover crops reduce nitrate leaching in agroecosystems: A global meta-analysis. *Journal of Environmental Quality, 47*, 1400–1411.

White, C. M., & Weil, R. R. (2011). Forage radish cover crops increase soil test phosphorus surrounding holes created by radish taproots. *Soil Science Society of America Journal, 75*, 121–130.

7

Soil Gas Emissions

7.1 Carbon and Nitrogen Emissions

Significant amounts of C and N as CO_2, N_2O, and CH_4 emissions are lost to the atmosphere from intensively cultivated lands (USEPA, 2019). Reducing such emissions is critical for maintaining, restoring, and accumulating C in the soil, improving N use efficiency and soil fertility, and enhancing overall soil ecosystem services. One of the potential strategies that can contribute to C and N cycling is the introduction of cover crops into current cropping systems. Growing cover crops can capture CO_2 via photosynthesis, increase biomass C input, and potentially restore C in the soil. Similarly, legume cover crops can fix N from the atmosphere, produce available N for crops, and reduce the amount of inorganic N fertilizer needed. Growing plants year-round can allow continuous capture of CO_2 and N in croplands. However, it is also important to note that cover crops can emit CO_2 via respiration (e.g., leaves, roots), which may reduce the total amount of C that cover crops can accumulate in the soil. An understanding of the whole C and N cycle and budget under cover crops on an annual basis and across multiple years can help to better evaluate how cover crops impact gains and losses of C and N.

While cover crop impacts on soil erosion, soil properties, soil fertility, and other services are relatively well known, the magnitude of cover crop impacts on soil CO_2, N_2O, and CH_4 emissions and the factors that may affect such emissions have not been discussed to the same extent. Cover crops are expected to alter soil C and N dynamics by adding aboveground and belowground biomass as well as by altering soil physical, chemical, and biological processes. For instance, cover crop biomass contains approximately 45% C and 1–4% total N content, depending on cover crop species. Thus, addition of biomass with cover crops can directly alter

---

Cover Crops and Soil Ecosystem Services, First Edition. Humberto Blanco.
© 2023 American Society of Agronomy, Inc. / Crop Science Society of America, Inc.
/ Soil Science Society of America, Inc. Published 2023 by John Wiley & Sons, Inc.

soil organic C and N concentrations and thus soil gas emissions. This chapter discusses how introduction of cover crops affects CO_2, N_2O, and CH_4 emissions and the potential factors that may drive such emissions.

7.2　Carbon Dioxide

Available research data and reviews indicate cover crops generally increase CO_2 emissions (Ruis et al., 2018; Muhammad et al., 2019). Cover crops can increase CO_2 emissions through combined effects of plant respiration, cover crop residue decomposition, and increased soil organic C concentration, microbial activity, and soil macroporosity (Muhammad et al., 2019). It is important to note that the available studies commonly included cover crop plants within the gas chamber during measurement and thus report data on the combined response of cover crop and soil respiration. Separating plant respiration from soil respiration during the measurement of CO_2 emissions can be difficult, but the separation would be useful to fully understand the extent to which cover crops increase CO_2 emissions from the soil.

Available reviews show that the increase in CO_2 emissions with cover crops can be as high as 90% relative to no cover crops (Ruis et al., 2018; Muhammad et al., 2019). Increased microbial activity with cover crops accelerates mineralization of organic matter when soil moisture and temperature conditions are favorable. Also, cover crops increase the concentration of different soil organic C fractions including stable and labile C (Beehler et al., 2017). In the short term, cover crops could alter labile C or particulate organic matter concentration more than total organic C in some cases (Crespo et al., 2021). The labile C serves as substrate for decomposers and is susceptible to rapid losses to the atmosphere as CO_2. The impact of cover crops on CO_2 emissions is highly variable, depending mainly on cover crop growing and non-growing season, biomass production, and climate (Table 7.1).

The loss of C with cover crops in the form of CO_2 may seem contradictory when one of the main reasons for adding cover crops is to gain C. However, it is important to discuss not all C gained from cover crops is lost. While cover crops can increase CO_2 emissions, cover crops could still accumulate C in the soil after a full accounting of C gains in the soil profile and C losses after cover crop introduction. In other words, a fraction of C input from cover crops is lost as emissions, but a significant portion can be gained, particularly through cover crop root biomass C input. Thus, the overall C budget after cover crop introduction can be greater compared to no cover crop introduction. Also, the potential benefits of cover crops on soil ecosystem services including erosion control, reduced leaching, and others could outweigh the losses of C via respiration. Overall, cover crops generally increase CO_2 emissions but most of the CO_2 emissions occur when cover crops are growing.

Table 7.1 Potential Impacts of Different Factors on CO_2 and N_2O Emissions after Cover Crop Introduction

Factors	Response of CO_2 emissions	Response of N_2O emissions
Cover crop species	Increase with all cover crop species	Non-legume species reduce; Legume species increase
Cover crop biomass	Increase as biomass amount increases, particularly above $2\,Mg\,ha^{-1}$	Increase as biomass amount increases, particularly above $5\,Mg\,ha^{-1}$
Cover crop biomass C to N ratio	Decrease as C to N ratio increases although this can be temporary until residues start to decompose	
Tillage system	Increase as tillage intensity increases	
N fertilization	Increase as N rate increases	
Soil texture	Increase from silt loam and sandy loam to clayey and silty clay loam	
Precipitation	Increases as precipitation increases	

Notes: Basche et al. (2014); Ruis et al. (2018); Muhammad et al. (2019).

7.3 Nitrous Oxide

Approximately 75% of N_2O lost to the atmosphere from agricultural lands in the U.S. are due to N fertilizers (USEPA, 2019). Nitrogen use efficiency is lower than P and K use efficiency, attributed to high N mobility and high N fertilizer application. Crops often use only approximately 50% of the N applied during the growing season, which leaves 50% of N prone to losses via leaching, emissions, and erosion. Most N is lost during the fallow period when crops are absent and soil and weather conditions are favorable for N losses. Growing cover crops during this period can be a strategy to capture available N and thus reduce N_2O emissions. The higher the concentration of ammonium and nitrates in the soil from inorganic fertilizers and manure, the greater the potential for nitrification and denitrification. Cover crops can immobilize N, but increased nitrification and denitrification by microbial activity may partly offset N immobilization by cover crops. Indeed, labile C from cover crops is an essential substrate for microbes responsible for nitrification and denitrification.

While cover crops generally increase CO_2 emissions, their impacts on N_2O emissions appear to be mixed, depending on cover crop species and other factors (Muhammad et al., 2019). For example, a global review by Basche et al. (2014) concluded that cover crops increased N_2O emissions in 60% of studies and reduced N_2O emissions in 40% of studies. In turn, Ruis et al. (2018) found that cover crops increased N_2O emissions in 42% of studies, had no effect in 42%, and decreased in 16%. The increase in N_2O emissions can range from 34 to 76% (Ruis et al., 2018). When cover crops reduce N_2O emissions, such reduction can be due to N immobilization in plant tissues and use of water, which reduces denitrification. The inconsistencies in the magnitude of cover crop impacts on N_2O emissions can be attributed to numerous factors, particularly cover crops species, biomass amount, and C to N ratio as discussed in the next sections (Table 7.1). Overall, available data indicate cover crops can have mixed effects on N_2O emissions but can increase CO_2 emissions.

7.4 Methane

Cover crops impact emissions of C in the form of CO_2 more than in the form of CH_4 (Wegner et al., 2018). Cover crops can increase, decrease, or have no effect on CH_4 emissions. Their effect may depend on the cropping system. Cover crops often increase CH_4 emissions in wet or flooded soils such as under rice paddies where anaerobic respiration produces methane via metabolism (methanogenesis; Li et al., 2021). Under upland cropping systems (e.g., corn, soybean, wheat) with aerobic conditions, cover crop impact on CH_4 emissions can be small or negligible (Wegner et al., 2018). Incorporation of cover crop residues into the soil could increase emissions of CH_4 and other gases as residue-soil water contact is favored in deeper layers for methane production. In China, Li et al. (2021) found that incorporation of winter vetch and ryegrass cover crop residues into the soil significantly increased CH_4 emissions under continuous rice cropping system. Overall, CH_4 emissions can be less responsive to cover crop introduction than CO_2 and N_2O emissions although multi-year experimental data on cover crops and CH_4 emissions are few.

7.5 Factors Affecting Soil Gas Emissions

A number of factors affect cover crop impacts on soil gas emissions including cover crop species, cover crop biomass production, cover crop management, N fertilization, tillage system, and others. For example, cover crop biomass production determines the extent of changes in soil water content and soil temperature, which directly affect soil gas emissions. An increase in soil water content and soil temperature is significantly and positively correlated with soil gas emissions (Ruis et al., 2018). Managing the factors above can be essential to reduce losses of C and N in the form of gases (Figure 7.1).

7.5 Factors Affecting Soil Gas Emissions

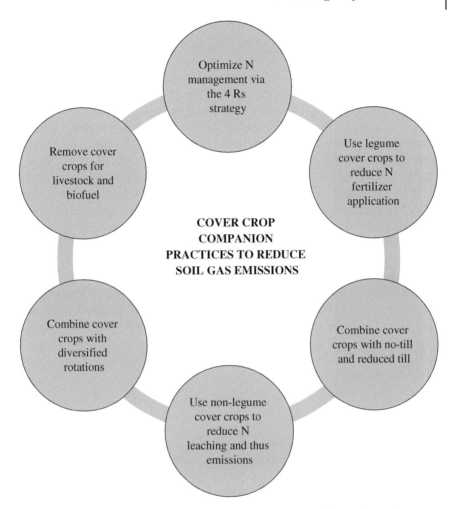

Figure 7.1 Some strategies that can enhance cover crop potential to reduce soil gas emissions.

7.5.1 Cover Crop Species

Emissions of some soil gases can depend on cover crop species. Emissions of CO_2 and CH_4 do not often vary among cover crop species, but N_2O emissions can depend on cover crop species. Research indicates, in general, non-legume cover crops reduce N_2O emissions, whereas legume cover crops increase N_2O emissions (Muhammad et al., 2019). Particularly, grass cover crops can reduce N_2O emissions more than legume and brassica cover crops (Basche et al., 2014). Grass cover crops are not only effective at reducing N_2O emissions but also at reducing nitrate leaching. Thus, grass cover crops can indirectly reduce N_2O emissions by reducing N losses via leaching. Under grass cover crop, more N is immobilized and thus less N is available for conversion to N_2O.

The contrasting effects of cover crop species on N_2O emissions are attributed to the mode of N acquisition and differences in biomass C to N ratio and mineralization rates (Muhammad et al., 2019). The latter factor determines the decomposition rate of cover crop biomass. Legume cover crop residues, due to their higher N concentration or lower C to N ratio decompose much faster than non-legume cover crops, releasing significant amounts of N as N_2O (Basche et al., 2014). Legume cover crops fix N from the atmosphere and can rapidly release N as N_2O emissions, whereas non-legume cover crops scavenge and immobilize N from the soil. Cover crop residue decomposition and N_2O emissions can also vary within the same cover crop species, depending on the cover crop growing period (Muhammad et al., 2019). Early-terminated cover crops have lower C to N ratio and can thus decompose faster, potentially releasing more CO_2 than late-terminated cover crops.

Grass cover crops can contribute to the management of excess N in croplands. Grass cover crops use excess N for their growth, so they can reduce N losses following establishment. Thus, integrating non-legume cover crops into current cropping systems can complement other equally if not more important N management practices including optimization of N fertilizer application and use of no-till and complex crop rotations.

7.5.2 Biomass Production

Cover crop biomass production can be another important determinant of changes in soil gas emissions. For example, emissions of N_2O could be more sensitive to cover crop species than to cover crop biomass production (Ruis et al., 2018). Soil CO_2 emissions increase but N_2O emissions decrease as cover crop biomass production increases. According to a global review by Muhammad et al. (2019), differences in cover crop biomass production explained 63% of variability in the increase in CO_2 emissions and 55% in the decrease in N_2O emissions. Also, cover crops may not release significant amounts of gases unless biomass production exceeds a threshold level. The estimated threshold level of cover crop biomass production that can significantly increase gas emissions is approximately $2\,Mg\,ha^{-1}$ for CO_2 and approximately $5.0\,Mg\,ha^{-1}$ for N_2O emissions (Muhammad et al., 2019; Ruis et al., 2018). The threshold biomass level for N_2O emissions can be larger than for CO_2 emissions, indicating CO_2 emissions can be more sensitive to an increase in biomass production than N_2O emissions.

Data on soil gas emissions from different levels of cover crop biomass production under the same experiment can be important to establish the threshold levels of biomass that significantly increase soil gas emissions. In a study in the western U.S. Corn Belt, under rainfed and irrigated no-till continuous corn systems, winter rye cover crop produced, on average, $0.94\,Mg\,ha^{-1}$ when terminated approximately 30 days before corn planting and $3.6\,Mg\,ha^{-1}$ of biomass when terminated at corn planting, and thus winter rye differently affected soil gas emissions (Ruis et al., 2018). Because the amount of biomass produced under the late-terminated cover crop ($3.6\,Mg\,ha^{-1}$) was above the threshold biomass level, late-terminated cover crop increased CO_2 emissions during spring relative to early-terminated cover crop in which biomass production ($0.94\,Mg\,ha^{-1}$) was below the threshold biomass level. Winter cover crops planted late in fall and terminated early in spring in temperate regions often produce low amounts of biomass and can thus result in low CO_2 emissions.

Late-terminated cover crops may not only increase CO_2 emissions by increasing biomass production but also by improving soil properties such as microbial biomass and soil C concentration compared with early-terminated cover crops. However, increases in CO_2 emissions with late termination of cover crops are often confined to the cover crop growing period because most CO_2 emissions occur via respiration. Also, research indicates that cover crop mixes may not differ from single species in their effects on CO_2 emissions as the amount of biomass produced between mixes and monocultures is often similar (Florence & McGuire, 2020). However, a mix of non-legume and legume cover crops could reduce N_2O emissions more than a mix of legume cover crops due to the lower C to N ratio and higher mineralization rate under legume cover crops.

7.5.3 Nitrogen Fertilization

One of the top drivers of N_2O emissions from cover crops can be the amount of N fertilizer added to the main crops. Sources of N include inorganic fertilizers, animal manure, and green manure although inorganic N fertilizer is the most common source of N. Emissions of N_2O from cover crops can increase with an increase in N fertilizer rate as more N is available for ammonification and nitrification. Indeed, N fertilizer rate can be a sensitive predictor of N_2O emissions from cover crops (Han et al., 2017). The relationship may not be, however, linear as N fertilizer rate can interact in their effects on N_2O emissions with cover crop species, fertilizer management (e.g., type, fertilizer placement), and other factors. In a review, Basche et al. (2014) discussed that N_2O emissions from non-legume cover crops increased as N rate increased.

Data on soil gas emissions from the same cover crop experiment managed with different levels of N fertilizer applied are needed to better understand the performance of cover crop for reducing N_2O emissions under different N fertilizer management practices. If cover crops are more efficient than row crops for utilizing the N applied, then N_2O emissions from fertilized cover crops may be minimal compared with non-fertilized cover crops although data to support this conclusion are presently limited. Because N_2O emissions are sensitive to N rate, embracing the 4Rs rules of fertilizer management in conjunction with other improved practices is recommended to reduce N losses (Figure 7.1).

7.5.4 Tillage and Cropping System

As tillage intensity increases, losses of C and N to the atmosphere often increase due to soil exposure and accelerated decomposition of organic matter. Emissions of CO_2 normally increase in this order: Moldboard plow > Chisel plow > Reduced till > No-till. Thus, tillage method can influence cover crop effect on soil gas emissions. Tilling cover crop residues into the soil often increases soil gas emissions relative to cover crop residues left on the soil surface. Incorporation of cover crops into the soil increases residue contact with soil and microbes, which accelerates residue decomposition (Ferrara et al., 2020). In a three-year tillage and cover crop study on a Haplustalf in Italy, N_2O emissions in no-till were 40–55% lower than in conventionally tilled soils, but within no-till soils, N_2O emissions were 20–36% lower with rye than with hairy vetch cover crops (Fiorini et al., 2020). Thus, no-till with grass cover crops can have lower N_2O emissions than no-till with legume cover crops.

Cover crops combined with no-till management can be a better option for managing N_2O emissions than when combined with conventionally tilled systems. For instance, tilling cover crops into the soil can release N_2O and potentially undo the reduction in N_2O emissions with cover crops. Cover crop residues left on the surface of no-till soils can reduce CO_2 emissions by reducing soil temperature. Soils under no-till cover crop residues are cooler in spring, when CO_2 emissions are often the highest, compared with soils with cover crop residue incorporated. While adding non-legume cover crops to no-till systems reduces N_2O emissions more than legume cover crops, adding legume cover crops can also be beneficial for reducing N_2O emission potential by fixing atmospheric N and thus reducing the amount of N fertilizer needed (Figure 7.1).

The effectiveness of cover crops for reducing soil gas emissions is also affected by cropping system or rotation phase. In a study in the U.S. Midwest, decomposing winter rye cover crop residues tended to increase N_2O emissions under the corn phase but tended to reduce N_2O emissions under the soybean phase although on an annual basis, cover crops did not affect N_2O emissions compared with no cover crops (Parkin et al., 2016). Thus, analyzing cover crop effects during years when a crop is fertilized (e.g., corn) and not fertilized (e.g., soybean) with N can provide a better understanding of cover crop effects on soil gas emissions.

Also, no-till cover crops combined with crop rotations can be a better option than when combined with monocultures such as corn. In Illinois, U.S., Behnke and Villamil (2019) reported that corn–soybean rotation reduced N_2O emissions by 35% (2 kg N ha^{-1} yr^{-1}) and increased corn yields compared with continuous corn or continuous soybean, but corn–soybean was no better than corn–soybean–wheat rotation. Thus, adding cover crops to corn–soybean rotation could further lower soil gas emissions, particularly N_2O emissions, while maintaining or increasing crop yields relative to corn monocultures, which require high rates of N fertilization.

Cover crops can be a strategy to reduce N_2O emissions, especially in cropping systems with long fallow periods when most N losses occur. For instance, in temperate regions, the fallow period between harvest and planting of crops such as corn or soybean is approximately seven months. The soil is thus without growing crops for approximately 60% of the year and during seasons when losses of C and N are more of a concern than during the main crop growing season.

7.5.5 Measurement Time

Cover crop impacts on soil gas emissions can vary between growing and non-growing cover crop seasons (Foltz et al., 2021). During the growing period, cover crops primarily emit CO_2 from plant respiration, but, after termination, they can release CO_2 from residue decomposition, which is high within two months after termination. Also, while non-legume cover crops often reduce N_2O emissions during the growing period, they could increase N_2O emissions after termination from residue decomposition (Basche et al., 2014; Foltz et al., 2021). Cover crop residues can also increase denitrification and thus N_2O emissions, depending on the precipitation amount. Denitrification can be high during winter and early spring when freeze–thaw cycles (temperate regions) are frequent and soil water content is high.

Cover crop residues often rapidly alter soil water content, labile C, and microbial processes, which can thus affect soil gas emissions in the short term. In the Northern Great Plains, U.S., after termination, non-legume cover crops including fall rye and oilseed radish increased N_2O emissions compared with no cover crops (Thomas et al., 2017). Most studies have measured gas emissions only for one year or only during cover crop growing season. Multi-year and multi-season monitoring of gas emissions within the same location can allow a better understanding of how emissions vary with season and year due to the potential variability in cover crop biomass production and slow changes in soil properties after cover crop introduction.

7.5.6 Soil Texture and Climate

Differences in soil properties and climate can directly affect the extent of soil gas emissions. Emissions of CO_2 with cover crops can increase from silt loam to sandy loam soils due to increased macroporosity and air-filled porosity and enhanced turnover of soil organic matter in coarse-textured soils (Muhammad et al., 2019). However, N_2O emissions can decrease from clay loam to sandy loam, which can be attributed to lower water content and denitrification in sandy soils. Tillage systems could interact with soil texture in their effects on soil gas emissions. For instance, no-till may increase N_2O emissions on poorly drained (anaerobic conditions), clayey, and compacted soils due to reduced soil porosity but may reduce or have no effect on N_2O emissions in medium-textured, well-aggregated, and well-drained soils.

Emissions of CO_2 can also be greater in regions with high than low annual precipitation and air temperatures. A review found that an increase in air temperature can explain 22% of the variability in N_2O emissions (Muhammad et al., 2019). Soil gas emissions increase with the combined increase in soil water content and soil temperature and often follow this order: Spring > Summer > Fall > Winter. Emissions can also vary with topography or landscape positions because soil C accumulation, erosion risks, and soil hydrology in addition to crop yield and biomass input often differ among landscape positions within the same field, especially in undulating fields. Such differences can differently alter soil gas emissions among topographical positions. In a humid continental climate, Nguyen and Kravchenko (2021) found cover crops including cereal rye and mixtures interacted with three landscape positions (summit, slope, and depression) in their effects on CO_2 emissions. They found cover crops significantly affected CO_2 emissions only in the summit position and not in other positions. In sum, while changes in soil water content and soil temperature are the main determinants of soil gas emissions, differences in topographic positions, soil texture, and initial soil C can also affect the extent of cover crop effects on soil gas emissions.

7.6 Summary

Reducing losses of soil C and N as gas emissions is essential for maintaining or restoring soil C, soil fertility, and other soil ecosystem services. Available experimental data provide important insights into how cover crops affect soil gas emissions. In general, cover crops increase CO_2 emissions and have limited or no effect on CH_4 emissions. Cover crop biomass quantity and quality are the main driver of soil gas emissions. For example, cover crop effect on N_2O emissions appears to depend on cover crop species.

Non-legume cover crops often reduce N_2O emissions, but legume cover crops can increase N_2O emissions. Non-legume cover crops take up and immobilize N from the soil unlike legume cover crops, which fix N from the atmosphere. Non-legume cover crops reduce N availability and water content, thereby enhancing no-till ability to reduce N_2O emissions. Emissions of CO_2 increase while N_2O emissions decrease with an increase in cover crop biomass production. It is well known that the amount of biomass produced determines changes in soil water content and soil temperature, which directly affect soil gas emissions.

Incorporation of cover crops into the soil as green manure can increase soil gas emissions by increasing organic matter turnover through soil disturbance relative to cover crop residues left on the soil surface. Soil texture can have variable effects on CO_2, but N_2O emissions often decrease from fine- to coarse-textured soils although more research is needed to elucidate how soil particle-size distribution affects soil gas emissions. Overall, fields with cover crops can emit more CO_2 than those without cover crops, but cover crops can still sequester more C in the soil than fields without cover crops due to additional biomass C input into the soil.

References

Basche, A. D., Miguez, F. E., Kaspar, T. C., & Castellano, M. J. (2014). Do cover crops increase or decrease nitrous oxide emissions? A meta-analysis. *Journal of Soil and Water Conservation, 69*, 471–482.

Beehler, J., Fry, J., Negassa, W., & Kravchenko, A. (2017). Impact of cover crop on soil carbon accrual in topographically diverse terrain. *Journal of Soil and Water Conservation, 72*, 272–279.

Behnke, G. D., & Villamil, M. B. (2019). Cover crop rotations affect greenhouse gas emissions and crop production in Illinois, USA. *Field Crops Research, 241*, 107580.

Crespo, C., Wyngaard, N., Rozas, H. S., Studdert, G., Barraco, M., Gudelj, V., Barbagelata, P., & Barbieri, P. (2021). Effect of the intensification of cropping sequences on soil organic carbon and its stratification ratio in contrasting environments. *Catena, 200*, 105145.

Ferrara, R. M., Martinelli, N., & Rana, G. (2020). CO_2 and H_2O fluxes due to green manuring under Mediterranean conditions. *The Italian Journal of Agrometeorology, 2*, 45–53.

Fiorini, A., Maris, S. C., Abalos, D., Amaducci, S., & Tabaglio, V. (2020). Combining no-till with rye (*Secale cereale* L.) cover crop mitigates nitrous oxide emissions without decreasing yield. *Soil and Tillage Research, 196*, 104442.

Florence, A. M., & McGuire, A. M. (2020). Do diverse cover crop mixtures perform better than monocultures? A systematic review. *Agronomy Journal, 112*, 3513–3534.

Foltz, M. E., Kent, A. D., Koloutsou-Vakakis, S., & Zilles, J. L. (2021). Influence of rye cover cropping on denitrification potential and year-round field N_2O emissions. *Science of the Total Environment, 765*, 144295.

Han, Z., Walter, M. T., & Drinkwater, L. E. (2017). N_2O emissions from grain cropping systems: A meta-analysis of the impacts of fertilizer-based and ecologically-based nutrient management strategies. *Nutrient Cycling in Agroecosystems, 107*, 335–355.

Li, J., Wang, S., Shi, Y. L., Zhang, L. L., & Wu, Z. J. (2021). Do fallow season cover crops increase N_2O or CH_4 emission from paddy soils in the mono-rice cropping system? *Agronomy*Basel, *11*, 199.

Muhammad, I., Sainju, U. M., Zhao, F., Khan, A., Ghimire, R., & Fu, X. (2019). Regulation of soil CO_2 and N_2O emissions by cover crops: A meta-analysis. *Soil and Tillage Research, 192*, 103–112.

Nguyen, L. T. T., & Kravchenko, A. N. (2021). Effects of cover crops on soil CO_2 and N_2O emissions across topographically diverse agricultural landscapes in corn-soybean-wheat organic transition. *European Journal of Agronomy, 122*, 126189.

Parkin, T. B., Kaspar, T. C., Jaynes, D. B., & Moorman, T. B. (2016). Rye cover crop effects on direct and indirect nitrous oxide emissions. *Soil Science Society of America Journal, 80*, 1551–1559.

Ruis, S., Blanco-Canqui, H., Ferguson, R. B., & Jasa, P. (2018). Impacts of early and late-terminated cover crops on gas fluxes. *Journal of Environmental Quality, 47*, 1426–1435.

Thomas, B. W., Hao, X., Larney, F. J., Goyer, C., Chantigny, M. H., & Charles, A. (2017). Non-legume cover crops can increase non-growing season nitrous oxide emissions. *Soil Science Society of America Journal, 81*, 189–199.

United States Environmental Protection Agency (USEPA). (2019). Overview of greenhouse gases. https://www.epa.gov/ghgemissions/overview-greenhouse-gases

Wegner, B. R., Chalise, K. S., Singh, S., Lai, L., Abagandura, G. O., Kumar, S., Osborne, S. L., Lehman, R. M., & Jagadamma, S. (2018). Response of soil surface greenhouse gas fluxes to crop residue removal and cover crops under a corn–soybean rotation. *Journal of Environmental Quality, 47*, 1146–1154.

8

Carbon Sequestration

8.1 The Need for Carbon Sequestration

Restoring the soil C lost due to intensive cultivation is imperative to improve the declining soil ecosystem services from agricultural lands. It is long recognized that soil C is a key driver and regulator of physical, chemical, and biological processes and properties of the soil. It determines the ability of the soil to function and deliver all the supporting, regulating, and provisioning services. For example, a decline in soil organic C levels can adversely affect the ability of the soil to recycle nutrients, water, air, and produce food, fuel, feed, and fiber.

Recent estimates indicate that CO_2 concentration in the atmosphere is reaching 420 ppm (NOAA, 2021), which indicates an abundance of CO_2 in the atmosphere available for photosynthesis and thus biomass production. Identifying or developing improved management practices that use the abundant atmospheric CO_2 and thus bring the C back to soil is a high priority. Innovative management practices that turn agricultural soils into C sinks rather than C sources are needed. The introduction of cover crops can be one such practice. Because growing plants use CO_2 from the atmosphere for aboveground and belowground biomass production, growing cover crops can be a biological practice to recapture C back into the soil. Cover crops, particularly non-legume species, may not only sequester C but also use the extra N during fallow periods, which would otherwise be lost through leaching or emissions.

The introduction of cover crops could be a potential strategy to use the abundant CO_2 from the atmosphere, sequester soil C, and thus enhance soil ecosystem services. However, the extent of C sequestration with cover crops could depend on many factors including biomass production, climate, and initial soil C, among others. Also, permanence or stability of C input from cover crops in the soil profile is not well understood. Growing cover crops for sequestering C in the soil is generating interest, but the potential site-specificity and challenges with cover crop

Cover Crops and Soil Ecosystem Services, First Edition. Humberto Blanco.
© 2023 American Society of Agronomy, Inc. / Crop Science Society of America, Inc. / Soil Science Society of America, Inc. Published 2023 by John Wiley & Sons, Inc.

establishment and production under different climates require further exploration. This chapter discusses, based on published information, some of the remaining questions about the potential of cover crops to sequester C.

- Do cover crops sequester significant amounts of C?
- Do cover crops sequester C in both topsoil and subsoil?
- What are the driving factors of C sequestration with cover crops?
- What are the strategies that can enhance the potential of cover crops to sequester C?

8.2 Rates of Carbon Sequestration

The available literature reviews and meta-analyses of data indicate cover crops can sequester significant amounts of C. The review findings summarized in Figure 8.1 show cover crops can sequester C between 0.21 and 0.56 Mg ha^{-1} yr^{-1}, depending on the review. Some reviews synthesized C sequestration rates under cover crops across all climates (Jian et al., 2020; Poeplau & Don, 2015; Ruis & Blanco-Canqui, 2017), while others emphasized C sequestration in temperate (McClelland et al., 2021) or semiarid (Blanco-Canqui et al., 2022) regions. Carbon sequestration rates vary among the available reviews due to the amount of data included or climatic region in question, but, on a global scale when averaged across all individual research studies, cover crops have potential to sequester C in the soil.

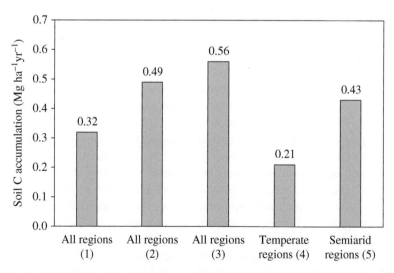

Figure 8.1 Published reviews on cover crop impacts on soil C sequestration. (1) Poeplau & Don, 2015; (2) Ruis & Blanco-Canqui, 2017; (3) Jian et al., 2020; (4) McClelland et al., 2021; (5) Blanco-Canqui et al., 2022.

The average C accumulation across all studies on regional or global scales summarized in Figure 8.1 should be, however, interpreted with caution. The average values (0.21–0.56 Mg ha^{-1} yr^{-1}) do not imply that cover crops will sequester the reported averages in all soils or fields after establishment. As an example, cover crops have mixed effects on soil C in the U.S. Midwest (Table 8.1). Cover crop biomass production, cover crop growing season, cover crop species, years after

Table 8.1 Some Case Studies on Cover Crops and Soil C from the U.S. Midwest

Location	Cropping system	Initial soil C (%)	Soil slope (%)	Cover crop	Years after adoption	Total soil C
[a]Kansas	Winter wheat–grain sorghum	1.37	<1	Summer (sunn hemp and soybean)	15	Increased in the 7.5 cm depth
[b]Illinois	Corn–soybean	na	6	Winter (hairy vetch and rye)	12	Increased in the 0 to 75 cm depth
[c]Illinois	Corn–soybean	na	6	Winter (hairy vetch and rye)	8	No change
[d]Nebraska	Continuous corn	1.83	<1	Winter rye	6	No change
[e]South Dakota	Corn–soybean	na	na	Lentils in winter and slender wheatgrass in summer	4	No change
[f]Nebraska	Continuous corn	1.68	<1	Winter rye	3	No change
[g]Iowa	Corn–soybean	3.89	<5	Winter rye and oat	3	No change
[h]Michigan	Corn–soybean	na	5.2	Spring red clover and winter rye	5	No change
[i]Indiana	Corn–soybean	na	na	Winter rye	4	No change

[a] Blanco-Canqui et al. (2011).
[b] Olson et al. (2014).
[c] Olson et al. (2010).
[d] Sindelar et al. (2019).
[e] Wegner et al. (2015).
[f] Blanco-Canqui et al. (2017).
[g] Kaspar et al. (2006).
[h] Ladoni et al. (2017).
[i] Rorick & Kladivko (2017).
na, Not available.

cover crop adoption, and other factors can affect cover crop impacts on soil C. Also, the reported C sequestration rates per year in Figure 8.1 do not mean that cover crops will linearly and indefinitely accumulate soil C at the above accumulation rates due to differences in C saturation among soils.

In some cases, cover crops can reduce soil organic C concentration (Tautges et al., 2019). This is probably because of the positive priming effect of fresh organic matter. Addition of fresh organic material with cover crops can trigger biological activity and accelerate the decomposition of new and existing organic matter, thereby reducing total organic C concentration. The numerous factors affecting C accumulation in the soil warrant assessment of cover crop potential for C sequestration on a local or site-specific basis.

8.3 Topsoil Versus Subsoil Carbon Sequestration

Most cover crop studies have measured soil C only for the upper 20 or 30 cm soil depth. Soil C measurement with depth is critical for a complete understanding of cover crop effects on C distribution in the whole soil profile. Shallow measurement of soil C cannot tell the whole story about cover crop effects for the soil profile. In a 19-year experiment under corn–tomato and wheat–fallow cropping systems in a Mediterranean climate, Tautges et al. (2019) found winter cover crops increased soil C stocks by 3.5% in the 0–30 cm depth but reduced it by 10.8% in the 30–200 cm depth, which resulted in an overall loss of C in the whole profile (2 m). Thus, measuring C only for the shallow depth can lead to erroneous conclusions.

The few studies that sampled soil to depths below 20 cm show that cover crops can preferentially accumulate C near the soil surface. One question is: How irreversible or how stable is the C accumulated near the soil surface? Increasing biomass C input through cover crops is not sufficient unless a large portion of C accumulated, if not all, is stored in stable forms. For example, a study in the central U.S. Great Plains found that cover crops accumulated C at a rate of 2.4–2.8 Mg ha^{-1} after five years (0.52 Mg ha^{-1} yr^{-1}); but, a year later after cessation of cover crop use, soil C stocks between plots with and without cover crops did not differ (Blanco-Canqui et al., 2013). This indicates that C accumulated with cover crops may not be stable. In other words, any gains in soil C with cover crops can be short-lived. This may be especially true in arid and semiarid regions where cover crop residues decompose rapidly after termination and any gains in soil C can be lost relative to cool climate regions.

Because cover crops are often paired with reduced till or no-till systems, the stratified accumulation of C under no-till cover crops is expected as no-till management often leads to stratification of C, nutrients, acidity, and other soil properties. One potential option to promote deep C sequestration in the soil profile could be planting cover crops with deep rooting depths (>0.5 m). However, because cover crops are grown only for a few months, cover crops do not often develop deep roots. Perennial cover crops can be a better option than annual cover crops for sequestering C (Poeplau et al., 2015). Perennial cover crops can mimic native perennial vegetation and could sequester more C in the soil and provide more soil ecosystem services than annual cover crops. However, how perennial cover crops can fit in current farming systems and impact crop production deserves further investigation.

Another option to transfer surface soil C to deeper depths can be the use of one-time or occasional tillage. While many consider that one-time tillage of long-term no-till soils could erase the benefits gained after many years of continuous no-till, research data show that one-time tillage in 5 or 10 years of long-term no-till does not undo the C gained with long-term no-till management (Blanco-Canqui & Wortmann, 2020; Dang et al., 2018). Instead, it can be an effective practice to more evenly distribute C in the soil profile, reduce C stratification, and possibly promote subsoil C sequestration. One-deep inversion of soil with tillage can place the C-enriched layer in deeper depths and bring the low-C layer to the surface for subsequent improvement with cover crops. The newly exposed C-depleted layer can more rapidly store C with cover crops than layers with C level near saturation. The buried C-enriched layer could increase stability and permanence of C in the subsoil (Alcantara et al., 2016).

It is often argued that soils, particularly in temperate regions, are near C saturation with limited room for C sequestration. While this argument may be true for surface layers, deeper layers can be less saturated with C than surface layers. Thus, subsurface layers can be an important sink for C in most soils. Also, buried C could facilitate soil aggregation in deeper depths where aggregation is often limited due to low C concentration. Stable aggregates in association with organo-mineral complexes can occlude and stabilize C. The soil in deep layers can have greater specific surface area and thus provide more opportunities for organo-mineral reactions than the soil in surface layers. Macroaggregates, which are more abundant near the soil surface than in the subsoil, have limited capacity to protect and stabilize C compared with microaggregates (Blanco-Canqui & Lal, 2004). Also, 2:1 clayey soils have high potential to trap and sequester C due to their higher specific surface area and negative charge compared with other clay minerals.

Emerging studies show deep tillage has the potential to increase C storage in deeper layers relative to current practices that confine C gains to near the soil surface. A meta-analysis of published data concluded that subsoiling rotated with long-term no-till management increased soil C stocks by approximately 9% in the 20–50 cm depth, particularly in arid regions with fine or medium-textured soils (Feng et al., 2020). Well-targeted and planned one-time tillage operation in 5 or

10 years may have a place in managing no-till systems, while potentially enhancing deep C storage. Note that one-time tillage may promote C burial while alleviating soil compaction, proliferation of herbicide resistant weeds, and other challenges associated with no-till management (Blanco-Canqui & Wortmann, 2020). One-time deep tillage may also enhance root penetration by breaking up compacted layers, which can further contribute to C accumulation in deeper depths.

While one-time tillage is argued here as a potential opportunity for deep C transfer in some cases, appropriate consideration of site-specific conditions is important. Ill-planned occasional tillage can increase soil erosion risks and accelerate C losses to the atmosphere unless cover crops or main crops are promptly established following occasional tillage to reduce risks of erosion. Also, while one-time or occasional tillage appears to be a potential remedy to reduce C stratification with long-term no-till cover crops, the fate and stability of the buried C for long-term C sequestration needs further research. Overall, cover crops, under current management scenarios, appear to accumulate C mostly near the soil surface, which warrants the consideration of potential practices that can allow transfer of surface C to deeper depths in the soil profile to boost long-term C stability and permanence.

8.4 Managing Carbon Sequestration

Among the factors that can affect soil C sequestration with cover crops include aboveground and belowground cover crop biomass production, cover crop species, years after cover crop adoption, tillage system, cropping systems, initial soil C level, soil texture, topography, and climate (Table 8.2). Interactions prevail among these factors. A single factor analysis does not fully capture the potential correlations among factors. Thus, the impact of a single factor cannot be evaluated separately from the impact of another factor. Approaches such as multiple regression analysis could better account for the interrelationships among factors relative to single factor analysis.

8.4.1 Biomass Production

A positive relationship between cover crop biomass production and soil C stocks is expected as biomass C input is indispensable for any gains in soil C. Indeed, cover crop biomass production is a sensitive predictor of changes in soil C (McClelland et al., 2021). Soil C increases as cover crop biomass amount increases, but low biomass production ($<1\,Mg\,ha^{-1}$) can have modest or no effects on soil C stocks. A review by McClelland et al. (2021) reported that biomass production above $7\,Mg\,ha^{-1}$ increased soil C stocks by 30% relative to $<7\,Mg$ biomass ha^{-1}.

Table 8.2 Potential Factors Affecting Soil C Sequestration with Cover Crops

Factors	Impact on soil carbon sequestration
Cover crop biomass production	An increase in biomass production increases soil C accumulation
Cover crop species	Cover crops accumulate soil C irrespective of species although legumes decompose more rapidly than grasses
Cover crop mixes	Mixes do not accumulate soil C more than single species
Crop planting and termination date	Early planting and late termination can promote C sequestration relative to late planting and early termination due to differences in biomass amount
Tillage system	Tillage systems may not affect cover crop effects on C sequestration although no-till is preferred over tilled systems
Initial soil C level	Low C soils can sequester more C than high C soils
Field topography	Sloping fields can accumulate more C than flat soils with high initial C levels
Soil texture	Fine-textured soils can accumulate more C than coarse-textured soils
Climate	Soil C stocks do not appear to differ among climatic regions

However, achieving this high amount of cover crop biomass production is not possible in all regions (Blanco-Canqui, 2022).

Furthermore, not all published studies report both soil C stocks and biomass amount, which limits a robust understanding of threshold level of cover crop biomass production needed to significantly change soil C stocks under different environments and cover crop management practices. Experiments with different cover crop termination dates to achieve different levels of aboveground and belowground biomass production within the same experiment can be ideal to establish minimum levels of biomass production for C sequestration for different soils and climatic conditions (Figure 8.2). As discussed in previous chapters, if cover crop biomass production is $<1\,\mathrm{Mg\,ha^{-1}}$, then improvement in soil properties and thus C accumulation under crops can be negligible (Figure 8.2). Also, root biomass production can be a more important determinant of soil C gains than aboveground biomass as most soil C originates from roots (Xu et al., 2021), but studies reporting both cover crop root biomass and soil C stocks are limited to study relationships between cover crop root biomass and soil C accrual. However, it is clear that an increase in root biomass production can lead to enhanced C sequestration as roots contain approximately 40% C. As indicated in Chapter 2, the amount of cover crop root biomass produced can be approximately 40–50% of the aboveground biomass produced.

Figure 8.2 Early-terminated cover crop (top) may not increase organic soil C concentration due to limited amount of biomass produced compared with late-terminated (at corn planting) cover crop (bottom) in cool temperate regions. Photo by H. Blanco.

8.4.2 Cover Crop Species and Mixes

Differences in cover crop species may have lesser effects on soil C stocks than differences in cover crop biomass production. Indeed, non-legume and legume cover crops under the same amount of biomass produced can have similar effects on soil C levels (Blanco-Canqui, 2022; McClelland et al., 2021; Poeplau & Don, 2015). However, turnover rates between legume and non-legume cover crop residues can differ due to differences in residue quality (C to N ratio). Residues of legume cover crops, which have low C to N ratio, can rapidly decompose, and increase labile C and soil microbial activity, but the turnover of microbial biomass can result in stable C (McClelland et al., 2021). Particularly, data on C to N ratio of roots from different cover crop species can be important to compare root-derived C stability between legume and non-legume cover crops. Differences in root C to N ratio between legume and non-legume could influence C turnover more than those in aboveground biomass C to N ratio as belowground biomass contributes more to stable C in the soil (Xu et al., 2021). More experimental data are needed to discern how different cover crop species contribute to soil C sequestration.

It is often hypothesized that diversifying cropping systems with diverse cover crop species can deliver multiple soil ecosystem services including soil C accumulation relative to single species. Thus, the question is: Do cover crop mixes really sequester soil C more than monocultures? Research data indicate the answer to this question is no at this point. A global review by Florence and McGuire (2020) found best-performing cover crop mixes did not accumulate more C in the soil than best-performing single species in most cases (88% comparisons) when mixes were compared against all their constituents grown as single species. In some cases, monoculture cover crops that produce more biomass than mixes can sequester soil C more than mixes.

The limited effect of mixes on soil C can be due to lack of differences in cover crop biomass production between mixes and monocultures (Florence & McGuire, 2020). Published data indicate that cover crop species mixes do not outperform monocultures for sequestering soil C. It is important to stress that comparing mixes only against all their constituents present as monocultures can be critical to draw valuable conclusions about mix effects. Many studies exist where not all species present in the mix were grown as monocultures. These studies do not allow a valid comparison on whether mixes outperform their counterparts when grown as single species. Because high-biomass producing single species can outperform mixes, planting single species can be more cost-effective. Overall, soil C sequestration between mixes and monoculture cover crops does not generally differ due to their similarities in biomass production.

8.4.3 Years after Cover Crop Adoption

Cover crops can have limited potential to sequester C in the short term (<5 yr). Direct addition of C to soil such as via application of large amounts of animal manure or biochar can rapidly increase soil C stocks, but accumulating C with growing plants, specifically cover crops, can take time, at least five years, if biomass production is consistently high enough. The latter depends on precipitation input and favorable temperature. The scale of changes in soil C after planting cover crops is often misunderstood. Any small changes in soil C from cover crops in the first few years can be hardly measurable due to both high variability in soil C changes in the field and limited sensitivity of current C assessment techniques. In Finland, a five-year assessment of C sequestration potential across 105 farms showed that, C farming practices, including cover crops, had limited potential to sequester large amounts of C under on-farm conditions (Mattila et al., 2022).

Furthermore, the potential linearity in soil C accumulation with years after cover crop adoption needs discussion. Soil C accumulation could linearly increase as time after cover crop introduction increases in the first few decades (e.g., 10–50 year). Poeplau and Don (2015) found that soil C accumulation was linearly correlated with time after cover crop introduction up to 54 years. The relationship between soil C accumulation and time may not be linear in the longer term (e.g., >50 yr) as soil becomes saturated with C. It is estimated that cover crops could accumulate significant amounts of C in the first 10–30 years and plateau or decrease after 100 or 150 years till a new equilibrium level of C is established (Jian et al., 2020; Poeplau & Don, 2015). Most of the available studies on cover crops and changes in soil C stock are short term (<5 yr). Thus, our understanding of how cover crop impacts C accumulation with time after establishment is still limited.

As an example, studies from the U.S. Midwest suggest that cover crops may increase soil C stocks only after 10 years (Table 8.1). One of the studies that monitored changes in soil C with time in a temperate soil in the eastern U.S. under the same experiment found that cover crops did not increase soil C stock after 8 years but increased it after 12 years in a corn–soybean system regardless of tillage system (Olson et al., 2010; Olson et al., 2014). Planting cover crops on the same field for many decades across different soils and climatic conditions can help to accurately understand how soil C accumulation changes with time and when soils approach C saturation.

8.4.4 Initial Soil Carbon Level

One of the factors that is not much discussed when assessing soil C sequestration by cover crops is initial or baseline soil C level. Indeed, cover crops are often adopted without accounting for the initial soil C level in the field. Thus, information on initial soil C level is often absent in literature, which makes evaluation of C sequestration rates difficult (Jian et al., 2020). Most of the available studies have computed C sequestration rates simply by dividing the difference in soil C stock between cover crop and no cover crop treatments over years after cover crop introduction. Information on initial C level is critical to accurately quantify whether cover crops are accumulating C with time after introduction or simply maintaining soil C levels relative to no cover crops.

First, soils with low initial C level can sequester more C relative to soils with high initial C level. Because high-C soils are closer to C saturation, they can be slower to respond to cover crop introduction or have less room for soil C accumulation than low-C soils. The examples of studies from the U.S. Midwest in Table 8.1 appear to corroborate that cover crops may not rapidly increase soil C if initial C level is high (>2% C). The time needed to reach C saturation after cover crop introduction depends on the initial C level provided that biomass production is appropriate. In a review, Poeplau and Don (2015) estimated that there was no sign of C saturation up to 54 years after cover crop introduction. This estimate will vary with soil and climate.

Second, it is often assumed that introduction of cover crops will increase soil C accumulation in all soils. However, while soil C level under cover crops can be higher than under no cover crops (Figure 8.1), soil C level in the whole field, in some cases, may remain unchanged or decline relative to pre-treatment C levels. In other words, in a few cases, even when cover crops are increasing soil C levels compared with no cover crops, the whole field may be losing C relative to pre-cover crop C levels. In this scenario, cover crops may not be increasing soil C but simply maintaining soil C levels or minimizing soil C losses. The latter scenario is also beneficial for soil C management but the system as a whole is not accumulating more C than before cover crop introduction. This consideration highlights the need for considering initial C level for cover crop management decisions.

8.4.5 Tillage Systems

It is often thought that frequent and intensive tillage could undermine the potential of cover crops to sequester C relative to no-till cover crops as tillage can trigger release of C by dispersing soil aggregates and exposing aggregate-occluded C to rapid oxidation. However, meta-analyses of available data report that soil C stocks between no-till and conventionally tilled soils do not generally differ (McClelland et al., 2021; Poeplau & Don, 2015). Thus, the available data appear to indicate that even tilled systems can benefit from cover crops and store similar amounts of C to no-till management. Annual tillage of cover crops may not always erase or cancel out the benefits from cover crops. The limited or no negative effect of tillage on C gains with cover crops appear to favor our earlier discussion that one-time inversion tillage of long-term no-till may actually enhance deep C sequestration.

It is critical to examine how different tillage systems impact cover crop effects on soil C for the whole soil profile. However, long-term cover crop studies assessing C sequestration under both no-till and tilled systems are rare. In Illinois, U.S., a 12-year study of no-till, chisel plow, and moldboard plow corn–soybean rotation with and without hairy vetch and rye cover crops found that cover crops sequestered 1.21 Mg C ha^{-1}yr^{-1} in no-till, 0.35 Mg C ha^{-1}yr^{-1} in chisel plow, and 0.55 Mg C ha^{-1}yr^{-1} in moldboard plow for the 0–75 cm soil depth (Olson et al., 2014). This individual study suggests combining cover crops with no-till could accumulate more C in the soil than when combined with tilled systems in some cases.

Differences in residence time of soil C in deeper depths among different tillage systems are still unclear. It is hypothesized that tilling cover crops into the soil may reduce soil C stocks in the upper 5 or 10 cm soil depth, but it may increase C stock in deeper depths compared with no-till. The deep C in tilled soils could have longer residence time or greater stability than shallow C in no-till soils. Overall, how the different tillage systems affect soil profile-C distribution and C residence time merits a site-specific and long-term evaluation.

8.4.6 Soil Texture

Soil-particle size distribution can affect soil C dynamics and storage after cover crop introduction. Soil C storage under cover crops generally increases with an increase in clay content and decrease in sand content (Jian et al., 2020). McClelland et al. (2021) concluded that soils with >20% clay content had 13% greater soil C stocks than those with <20% clay content. Clay particles can react and interact with soil C more than sand particles as clay particles have higher specific surface area than sand particles. Clay particles can also promote aggregation and form stable microaggregates and macroaggregates. Stable aggregates can protect C within aggregates and reduce access to microbes.

Because coarse-textured soils often have limited aggregation, C added by cover crops can be more prone to rapid losses such as through leaching as dissolved C. Soil C stocks in silt loams and other medium-textured soils with high silt content can be between clayey and sandy soils. Sandy soils also have lower water and nutrient holding capacity than fine-textured soils and thus may produce less biomass, which can reduce amount of C input. Also, the positive priming effect can be particularly significant in coarse-textured soils with initial low organic matter under warm climates, whereas negative priming often occurs in fine-textured soils under cool temperate climates. The positive priming effect can explain the limited potential of coarse-textured soils to accumulate or sequester large amounts of C in spite of high levels of cover crop biomass production (Santos et al., 2022). Overall, fine-textured soils can have higher efficiency for accumulating soil C than coarse-textured soils. It is important to note, however, that, in general, soil particle size and C storage relationships can be complex, depending on the clay type and mineralogy as well as climate. For example, in some cases, soil C accumulation can be related to both clay and silt content (Zinn et al., 2005) or to both clay and sand (Wang et al., 2013).

8.4.7 Topographic Characteristics

Most available data on cover crops and soil C stocks are from relatively flat or nearly-level soils (<3% slope). This is because most cover crop experiments are often established in small plots with homogeneous topographic conditions. However, soil C data from such ideal topographic conditions may not fully reflect the potential of cover crops to accumulate C under fields with undulating topography, which is typical in most croplands. Soils in nearly-flat topography often have higher initial soil C concentration and can thus be slower to respond to cover crop introduction compared with sloping soils. In the upper U.S. Midwest, Kaspar et al. (2006) found that soil C concentration decreased in an exponential pattern as soil slope increased from approximately 0% to 5%. Soil C concentration was approximately $50\,g\,kg^{-1}$ for soil slopes <0.5% and approximately $20\,g\,kg^{-1}$ for 4–5% soil slopes.

In sloping fields, backslope or shoulder positions often have lower soil C concentration, due to high erosion rates, than the footslope or toeslope positions. For instance, in 20 large corn–soybean–wheat rotation fields in Michigan, U.S., compared with no cover crop, rye and red clover cover crops established in three landscape positions (summit, slope, and depression) preferentially increased soil labile C concentration in slope and summit positions and not in the depression position in spite of higher cover crop biomass production in depression areas (Ladoni et al., 2017).

Targeting sloping fields or eroded landscape positions with cover crops can be potential strategy to sequester C relative to planting cover crops in low-lying field positions. In sloping soils, cover crops may not only increase soil C stocks by adding biomass C but also by reducing soil erosion and thus losses of C associated with sediment. Also, because sloping fields are more susceptible to degradation and have lower C levels than nearly-level soils, sloping fields can benefit more from cover crops than nearly-level soils. Overall, available data indicate cover crops can interact with landscape positions within the same field in their magnitude of effects on C sequestration. An accurate assessment of the potential of cover crops to sequester C at local and regional scales requires a quantification of differences in the amount of C sequestered for different topographic conditions within a field.

8.4.8 Climate

Cover crops could potentially store similar amounts of C in temperate and tropical climates under appropriate management. In a review, Poeplau and Don (2015) found change in soil C stock with cover crops between temperate and tropical climates did not differ. Another review concluded cover crops can increase soil C stock by, an average, of $0.43\,Mg\,ha^{-1}\,yr^{-1}$ in semiarid regions (Blanco-Canqui et al., 2022). This rate of accumulation in semiarid regions is similar to that reported for other regions (Figure 8.1). Although cover crops can reduce soil water and potentially reduce subsequent crop yields in some semiarid soils, available data suggest cover crops under appropriate management can accumulate significant amounts of C even in semiarid soils. This may lead to the consideration that the increased in soil C concentration with cover crops can improve the ability of the soil to retain water, potentially reducing or offsetting the amount of water used by cover crops in water-limited regions.

Cover crops can, on average, increase soil C stocks in all climatic regions, but it is critical to discuss the site-specificity of cover crop impacts within the same climatic region. For example, in cold temperate regions with high initial soil C concentration, such as those in the U.S. Midwest (Table 8.1), cover crops may not sequester C, particularly in the short term (<5year) and when biomass production is low. A review with focus on temperate regions discussed that cover crops can store more C in subtropical than in cool temperate regions due to higher precipitation and temperature, which favor cover crop biomass production in subtropical regions (McClelland et al., 2021). Soil characteristics (e.g., initial C level) and cover crop management in interaction with climate will determine whether cover crop accumulates C within a climatic region. Thus, site-specific characteristics can be better predictors of cover crop-induced changes in soil C stocks than climatic regions.

8.5 Cropping System Carbon Footprint

Literature indicates introduction of cover crops to current cropping systems has potential to reduce the overall C footprint from such systems by sequestering C compared with cropping systems without cover crops (Figure 8.1). Adopting cover crops is promising to lower the C footprint from agricultural lands, but it is important to consider the whole C cycle with cover crop introduction (Prechsl et al., 2017). Cover crops capture CO_2 from the atmosphere for biomass accrual, but, at the same time, C footprint through planting (e.g., fuel use), herbicide use for chemical termination, and others should be considered for C budgeting.

Also, in some cases, cover crops are irrigated in water-limited regions or fertilized to ensure successful establishment. Irrigation of cover crops can increase CO_2 emissions by using energy for pumping water and activating soil microbial processes and accelerating organic matter decomposition relative to rainfed cover crops (Follett, 2001). Similarly, fertilization of cover crops can increase CO_2 emissions through fertilizer production, transport, and application. However, irrigated and fertilized cover crops can produce more biomass and thus C input that non-irrigated and non-fertilized cover crops, which can potentially offset the increased CO_2 emissions with irrigation and fertilization. Additionally, cover crops during the growing season and after termination can emit CO_2 and other soil gases, which may reduce their C sequestration potential. In sum, all the C inputs and outputs need accounting through comprehensive life cycle analysis to establish the magnitude of cover crop impacts on C sequestration.

8.6 Strategies to Enhance Cover Crop Potential to Sequester Carbon

A number of possible strategies exist to enhance the amount of C that cover crops can accumulate in the soil. Any strategy that increases aboveground and belowground biomass production, discussed in Chapter 2, can concomitantly increase

Figure 8.3 Some potential strategies to promote C sequestration with cover crops.

potential of cover crops to sequester C. Increasing cover crop biomass production by extending the cover crop growing season and establishing cover crops in low C soils or eroded soils can be top strategies to increase C accumulation with cover crops (Figure 8.3). Increasing the length of cover crop growing season or window can directly increase the amount of biomass produced when soil moisture and temperature are not limiting. Some strategies to increase the cover crop growing window include interseeding cover crop, planting after corn silage or seed corn, planting after small grain (e.g., winter wheat) harvest in summer, planting after early-maturing crops, and others.

Particularly, increasing root biomass production can increase soil C accumulation as root-derived C is predominantly responsible for gains in stable soil C. Besides root biomass production and architecture, root chemistry and decomposition rates can affect soil C turnover and C sequestration (Xu et al., 2021). How the physical, chemical, and biological properties of different cover crop roots affect transformation and accrual of C is still under investigation. Also, because potential of no-till alone for sequestering C is questionable (Powlson et al., 2014), pairing high-biomass producing cover crops with no-till technology could have additives effects on increasing C sequestration. Especially, combining deep-rooted cover crops with no-till could favor C translocation to deeper horizons to ensure the permanence of C in the subsoil.

Another strategy to promote deep C sequestration with cover crops could be the use of one-time deep tillage in long-term no-till cover crop systems (Feng et al., 2020). As discussed earlier, a strategic deep tillage done once in 5 or 10 years may not erase all the benefits of long-term no-till cover crop systems for storing C and maintaining related ecosystem services (Blanco-Canqui & Wortmann, 2020). Indeed, deep inversion tillage can allow burial of C-enriched surface layers and transfer of the non-improved subsurface layer to the surface layer for subsequent improvement. Targeted-deep tillage can fracture soil compacted layers, reduce C and nutrient stratification, and provide other services (e.g, weed management) in long-term no-till systems. Overall, early cover crop planting, late cover crop termination, use of deep-rooted cover crop species, and strategic deep tillage can be some of the strategies to enhance soil C sequestration with no-till cover crops in agricultural lands.

8.7 Summary

Experimental data indicate that cover crops have significant potential to build up and sequester C in the soil. However, the extent to which cover crops can alter soil C stocks is highly site-specific, depending on cover crop biomass production, initial C level, soil texture, topography, climate, and other factors. Introducing cover crops may not result in rapid increases in soil C stocks if biomass production is low ($<1\,Mg\,ha^{-1}$), initial soil C is high ($>2\%$ C), and time after cover crop introduction is short (<5 years).

The high C sequestration rates observed in some cover crop studies may not be observed in all soils and climates, which cautions against overgeneralization of soil C storage benefits of cover crops. Also, cover crops appear to mainly accumulate C in the upper 20 cm soil depth and not in the subsoil although measured data for the whole soil profile C are limited. Growing deep-rooted cover crops and using one-time or occasional tillage (5 or 10 years) in long-term no-till cover crop systems can be strategies to transfer C to deeper depths and potentially promote long-term C permanence. One-time inversion tillage can bury the C-enriched layer and move the low-C layer to the surface for subsequent C enrichment.

Opportunities exist to sequester more C by planting cover crops in sloping than in flat portions of the field. Because one of the key drivers of soil C accumulation is biomass production, particularly belowground biomass production, extending the duration of the cover crop growing season (e.g., planting early, terminating late) can enhance biomass production and thus boost biomass C input.

References

Alcantara, V., Don, A., Well, R., & Nieder, R. (2016). Deep ploughing increases agricultural soil organic matter stocks. *Global Change Biology, 22*, 2939–2956.

Blanco-Canqui, H. (2022). Cover crops and carbon sequestration: Lessons from US studies. *Soil Science Society of America Journal, 86*, 501–519.

Blanco-Canqui, H., Holman, J. D., Schlegel, A. J., Tatarko, J., & Shaver, T. (2013). Replacing fallow with cover crops in a semiarid soil: Effects on soil properties. *Soil Science Society of America Journal, 77*, 1026–1034.

Blanco-Canqui, H., & Lal, R. (2004). Mechanisms of carbon sequestration in soil aggregates. *Critical Reviews in Plant Sciences, 23*, 481–504.

Blanco-Canqui, H., Mikha, M. M., Presley, D. R., & Claassen, M. M. (2011). Addition of cover crops enhances no-till potential for improving soil physical properties. *Soil Science Society of America Journal, 75*, 1471–1482.

Blanco-Canqui, H., Ruis, S., Holman, H., Creech, C., & Obour, A. (2022). Can cover crops improve soil ecosystem services in water-limited environments? A Review. *Soil Science Society of America Journal, 86*, 1–18.

Blanco-Canqui, H., Sindelar, M., & Wortmann, C. S. (2017). Aerial interseeded cover crop and corn residue harvest: Soil and crop impacts. *Agronomy Journal, 109*, 1344–1351.

Blanco-Canqui, H., & Wortmann, C. W. (2020). Does occasional tillage undo the ecosystem services gained with no-till? A review. *Soil Tillage Res., 198*, 104534.

Dang, Y. P., Balzer, A., Crawford, M., Rincon-Florez, V., Liu, H., Melland, A. R., Antille, D., Kodur, S., Bell, M. J., Whish, J. P. M., Lai, Y., Seymour, N., Costa Carvalhais, L., & Schenk, P. (2018). Strategic tillage in conservation agricultural systems of North-Eastern Australia: Why, where, when and how? *Environemental Science and Pollution Research, 25*, 1000–1015.

Feng, Q., An, C., Chen, Z., & Wang, Z. (2020). Can deep tillage enhance carbon sequestration in soils? A meta-analysis towards GHG mitigation and sustainable agricultural management. *Renewable and Sustainable Energy Reviews, 133*, 110293.

Florence, A. M., & McGuire, A. M. (2020). Do diverse cover crop mixtures perform better than monocultures? A systematic review. *Agronomy Journal, 112*, 3513–3534.

Follett, R. F. (2001). Soil management concepts and carbon sequestration in cropland soils. *Soil and Tillage Research, 61*, 77–92.

Jian, J., Du, X., Reiter, M. S., & Stewart, R. D. (2020). A meta-analysis of global cropland soil carbon changes due to cover cropping. *Soil Biology and Biochemistry, 143*, 107735.

Kaspar, T. C., Parkin, T. B., Jaynes, D. B., Cambardella, C. A., Meek, D. W., & Jung, Y. S. (2006). Examining changes in soil organic carbon with oat and rye cover crops using terrain covariates. *Soil Science Society of America Journal, 70*, 1168–1177.

Ladoni, M., Basir, A., Robertson, P. G., & Kravchenko, A. N. (2017). Scaling-up: Cover crops differentially influence soil carbon in agricultural fields with diverse topography. *Agriculture, Ecosystems and Environment, 225*, 93–103.

Mattila, T. J., Hagelberg, E., Söderlund, S., & Joona, J. (2022). How farmers approach soil carbon sequestration? Lessons learned from 105 carbon-farming plans. *Soil and Tillage Research, 215*, 105204.

McClelland, S. C., Paustian, K., & Schipanski, M. E. (2021). Management of cover crops in temperate climates influences soil organic carbon stocks - A meta-analysis. *Ecological Applications, 31*, e02278.

NOAA Global Monitoring Laboratory. (2021). Atmospheric CO2 at Mauna Loa Observatory. https://research.noaa.gov/article/ArtMID/587/ArticleID/2764/Coronavirus-response-barely-slows-rising-carbon-dioxide

Olson, K., Ebelhar, S. A., & Lang, J. M. (2014). Long-term effects of cover crops on crop yields, soil organic stocks and sequestration. *Open J. Soil Sci., 4*, 284–292.

Olson, K. R., Ebelhar, S. A., & Lang, J. M. (2010). Cover crop effects on crop yields and soil organic carbon content. *Soil Science, 175*, 89–98.

Poeplau, C., Aronsson, H., Myrbeck, A., & Kätterer, T. (2015). Effect of perennial ryegrass cover crop on soil organic carbon stocks in southern Sweden. *Geoderma Regional., 4*, 126–133.

Poeplau, C., & Don, A. (2015). Carbon sequestration in agricultural soils via cultivation of cover crops-A meta-analysis. *Agriculture, Ecosystems and Environment, 220*, 33–41.

Powlson, D. S., Stirling, C. M., Jat, M. L., Gerard, B. G., Palm, C. A., Sanchez, P. A., & Cassman, K. G. (2014). Limited potential of no-till agriculture for climate change mitigation. *Nature Climate Change, 4*, 678–683.

Prechsl, U. E., Wittwer, R., van der Heijden, M. G. A., Lüscher, G., Jeanneret, P., & Nemecek, T. (2017). Assessing the environmental impacts of cropping systems and cover crops: Life cycle assessment of FAST, a long-term arable farming field experiment. *Agricultural Systems, 157*, 39–50.

Rorick, J. D., & Kladivko, E. J. (2017). Cereal rye cover crop effects on soil carbon and physical properties in southeastern Indiana. *Journal of Soil and Water Conservation, 72*, 260–265.

Ruis, S. J., & Blanco-Canqui, H. (2017). Cover crops could offset crop residue removal effects on soil carbon and other properties: A review. *Agronomy Journal, 109,* 1785–1805.

Santos, G. G., Rosetto, S. C., Barbosa, R. S., Melo, N. B., Soares de Moura, M. C., Santos, D. P., Flores, R. A., & Collier, L. S. (2022). Are chemical properties of the soil influenced by cover crops in the Cerrado/caatinga ecotone? *Communications in Soil Science and Plant Analysis, 53,* 89–103.

Sindelar, M., Blanco-Canqui, H., Virginia, J., & Ferguson, R. (2019). Cover crops and corn residue removal: Impacts on soil hydraulic properties and their relationships with carbon. *Soil Science Society of America Journal, 83,* 221–231.

Tautges, N. E., Chiartas, J. L., Gaudin, A. C. M., & A.T. O'geen, I. Herrera, and K.M. Scow. (2019). Deep soil inventories reveal that impacts of cover crops and compost on soil carbon sequestration differ in surface and subsurface soils. *Global Change Biology, 25,* 3753–3766.

Wang, M. Y., Shi, X. Z., Yu, D. S., Xu, S. X., Tan, M. Z., Sun, W. X., & Zhao, Y. C. (2013). Regional differences in the effect of climate and soil texture on soil organic carbon. *Pedosphere, 23,* 799–807.

Wegner, B. R., Kumar, S., Osborne, S. L., Schumacher, T. E., Vahyala, I. E., & Eynarde, A. (2015). Soil response to corn residue removal and cover crops in eastern South Dakota. *Soil Science Society of America Journal, 79,* 1179.

Xu, H., Vandecasteele, B., De Neve, S., Boeckx, P., & Sleutel, S. (2021). Contribution of above- versus belowground C inputs of maize to soil organic carbon: Conclusions from a 13C/12C-resolved resampling campaign of Belgian croplands after two decades. *Geoderma, 383,* 114727.

Zinn, Y. L., Lal, R., & Resck, D. V. S. (2005). Texture and organic carbon relations described by a profile pedotransfer function for Brazilian Cerrado soils. *Geoderma, 127,* 168–173.

9

Soil Water

9.1 Soil Water Management

Managing soil water in agricultural lands, particularly under extreme weather events such as frequent droughts and floods, is an increasing challenge. Improving the ability of soils to capture and retain water in dry periods and rapidly release or drain water in wet periods is more important than before. Thus, one of the revolving questions is: Can cover crops be a strategy to manage soil water in agricultural lands under fluctuating weather conditions? How management practices such as cover crops affect soil water dynamics will directly affect soil resilience against extreme weather events. Cover crops alter soil hydrological cycle and processes. They affect soil water balance by affecting precipitation (e.g., rain, snowmelt) interception and partitioning into runoff, infiltration, and evaporation. A quantitative understanding of cover crop impacts on soil water balance is critical to manage soil water.

Cover crops need water, like any other plant, for their growth regardless of climatic regions or precipitation zones. Such water use by cover crops can be highly beneficial in wet soils or high precipitation regions to manage excess water, whereas it could have negative implications in water-limited environments by reducing available water for the subsequent crop. Often, cover crops are viewed as systems that will adversely affect soil water storage without much consideration of potential positive impacts on soil water dynamics and balance following cover crop termination (Nielsen et al., 2015). A research-based discussion on cover crop impacts on soil water in both high and low precipitation regions can help to better understand how cover crops can contribute to soil water management within each region. This chapter discusses how cover crops impact soil water dynamics as affected by different factors.

Cover Crops and Soil Ecosystem Services, First Edition. Humberto Blanco.
© 2023 American Society of Agronomy, Inc. / Crop Science Society of America, Inc.
/ Soil Science Society of America, Inc. Published 2023 by John Wiley & Sons, Inc.

9.2 High Precipitation Regions

Cover crops can be grown in high precipitation zones without major concerns of water depletion relative to regions where water availability is limited. Thus, some have suggested that cover crops are better suited to high than low precipitation regions (Kaspar & Singer, 2011; Unger & Vigil, 1998). In regions with high precipitation (>750 mm), use of water by cover crops can be beneficial to reduce excess water, particularly in poorly drained and clayey soils in wet and cool temperate regions. In the eastern Canadian prairies, a berseem clover cover crop increased soil warming in winter, enhanced soil thawing, and improved water drainage relative to fields without cover crops (Kahimba et al., 2008). The improved water drainage can potentially reduce flooding risks.

Further, water use by cover crops in spring when soils are often wet can allow earlier entry into the fields for planting and performing other field operations. In a three-year study in a temperate climate, faba bean cover crop and vetch-oat cover crop mix significantly reduced excess water at corn planting when cover crops produced as much as 10.5 Mg ha^{-1} of biomass (Alletto et al., 2022). Also, a seven-year study from a temperate climate reported that winter rye cover crop increased plant available water, on average, by 21–22% but had no significant effects on subsequent corn and soybean yields (Basche et al., 2016). These studies indicate that proper management of cover crops can improve soil water storage in non-water limited regions.

Systems with subsurface drainage tiles can also benefit from cover crops to reduce water drainage volume and thus reduce nitrate leaching. A meta-analysis of studies found cover crops can reduce water drainage by 10–40 mm by increasing evapotranspiration relative to no cover crops (Meyer et al., 2019). Growing cover crops longer can use more water and reduce excess water in high precipitation regions. The implications of such reduced water drainage with cover crops depend on climatic regions. While drainage reduction can be beneficial for reducing excess water and nitrate leaching in wet regions, it can have negative effects on water-limited regions by limiting groundwater recharge and reducing crop production.

9.3 Low Precipitation Regions

Concerns over potential water depletion have often limited cover crop adoption in water-limited regions (Figure 9.1; Holman et al., 2018; Nielsen et al., 2015; Unger & Vigil, 1998). However, the question is: What does research really indicate about cover crop effects on soil water in water-limited environments? A review of cover crops and soil water focused on water-limited environments across a total of 96 comparisons (cover crops versus no cover crops) found cover crops reduced

Figure 9.1 Cover crop growing during fallow period in winter wheat-fallow systems to improve soil ecosystems services in a water-limited region (457 mm mean annual precipitation) near Sidney, NE. Photo by H. Blanco.

field soil water content in only 50% of comparisons and had no effect in 50% (Blanco-Canqui et al., 2022). Cover crops reduced soil water content by an average of 23% in the 50% of comparisons. The reduction of soil water content in the 50% of cases is a concern as it could result in reduced subsequent crop yields, particularly when precipitation is low at crop establishment (Nielsen et al., 2015).

The potential reduction in soil water demands a careful consideration of cover crop management. Developing cover crop management practices that minimize water use by cover crops is much needed to reduce the adverse effects of cover crops on subsequent crop yields in water-limited regions. The adverse effects of cover crops on soil water will not only depend on precipitation amount but also on air temperature. Cover crops can reduce soil water more in warmer semiarid regions than in cool semiarid regions due to increased evaporation in warmer regions. For example, because evapotranspiration is higher in the southern U.S. Great Plains, cover crops can have more negative effects on soil water in this region than in northern U.S. Great Plains (Nielsen et al., 2015).

However, the lack of adverse effects of cover crops in 50% of cases indicates cover crops can maintain soil water in water-limited environments in some cases. It supports the notion cover crops, as expected, can reduce soil water content when growing, but after termination, cover crop residues can contribute to water recharge and storage between termination and main crop planting when precipitation is sufficient (Burke et al., 2021). While many argue against the use of cover crops in water-limited regions, research data indicate cover crops do not always result in reduced soil water content. A global meta-analysis by Wang et al. (2021) found that cover crops increased soil water storage by 6% to a depth of 30 cm in semiarid and humid regions but had minimal effects on arid and sub-humid regions compared with no cover crops. This indicates opportunities exist for cover crop introduction in water-limited regions under proper management (e.g., early termination; Burke et al., 2021).

A review of case studies shows that cover crop effects on soil water content can be positive, negative, and neutral (Table 9.1). Even in low precipitation regions (<750 mm), cover crops do not reduce soil water content in all cases (Table 9.1).

Table 9.1 Case Studies on Cover Crops and Soil Water Content Under Rainfed Conditions

Location	Precipitation (mm)	Cover crop impact on soil water content at termination
[a]South Carolina, U.S.	1198	No effect
[b]Missouri, U.S.	1026	Increased
[c]Iowa, U.S.	954	Increased
[d]Buenos Aires, Argentina	946	Increased
[e]Anhui, China	924	Reduced
[f]Gottingen, Germany	624	No effect
[g]Vojvodina, Serbia	610	Reduced
[h]Two sites in Nebraska, U.S.	500–512	No effect
[i]Western Kansas, western Nebraska, and eastern Colorado, U.S.	464–657	Reduced
[j]Madrid, Spain	376	Increased
[k]Toulouse, France	239–577	Reduced

[a] Payero et al. (2021).
[b] Rankoth et al. (2021).
[c] Basche et al. (2016).
[d] Villarreal et al. (2021).
[e] Zhang et al. (2022).
[f] Grunwald et al. (2022).
[g] Vujić et al. (2021).
[h] Rosa et al. (2021).
[i] Kelly et al. (2021).
[j] Gabriel et al. (2021).
[k] Alleto et al. (2022).

The mixed effects of cover crops on soil water suggest changes in soil water content with cover crops need to be assessed on a farm or location basis. Overall, experimental data indicate that cover crops may or may not reduce soil water for the next crop in water-limited regions. In approximately 50% of cases, cover crops could offset water use by the mechanisms discussed below.

9.4 Mechanisms of Soil Water Storage with Cover Crops

The specific mechanisms by which growing cover crops and cover crop residues can contribute to soil water content maintenance and offset the water use by cover crops in water-limited regions probably include (Figure 9.2):

- Well-established cover crops intercept and buffer raindrop impacts, and trap rainfall and snow, which can reduce surface runoff and contribute to water storage (Kaspar & Singer, 2011).

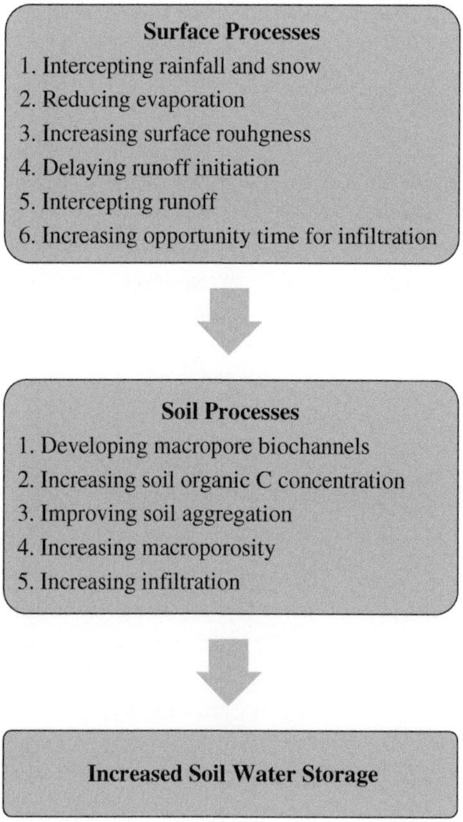

Surface Processes
1. Intercepting rainfall and snow
2. Reducing evaporation
3. Increasing surface rouhgness
4. Delaying runoff initiation
5. Intercepting runoff
6. Increasing opportunity time for infiltration

Figure 9.2 Cover crops contribute to precipitation capture and water storage by altering both surface processes and soil physical, chemical, and biological processes.

Soil Processes
1. Developing macropore biochannels
2. Increasing soil organic C concentration
3. Improving soil aggregation
4. Increasing macroporosity
5. Increasing infiltration

Increased Soil Water Storage

- Cover crop canopy and residues reduce soil surface sealing, delay runoff initiation, and increase opportunity time for water infiltration (Korucu et al., 2018).
- Soil-profile water recharge can occur if precipitation input is sufficient between cover crop termination and main crop planting (Wang et al., 2021). Cover crops with extensive and deep roots can open up biochannels and promote preferential flow of water to deeper depths in the profile.
- Cover crops can improve soil structural properties such as aggregate stability, macroporosity, and thus increase water infiltration and precipitation capture (Blanco-Canqui & Ruis, 2020). As discussed in an earlier chapter, cover crops can improve water infiltration in most cases.
- Cover crop residues left on the soil surface such as under no-till can insulate and protect the soil surface, reduce evaporation rates, and maintain soil water content relative to bare soils (Villarreal et al., 2021). Indeed, cover crops can alter soil water balance by changing soil temperature. Cover crops reduce daytime soil temperature and thus reduce evaporation rates. As discussed in Chapter 3, high-biomass producing cover crops can reduce daytime soil temperature by an average of 2 °C. The decrease in soil temperature conserves soil water (Blanco-Canqui et al., 2011).
- The positive effects of cover crop residues could last for extended periods of time after termination if residue produced is high. In the eastern and central U.S. Great Plains (mean annual precipitation of 874 mm), soils under late-maturing soybean and sunn hemp summer cover crop residues had 25% higher soil volumetric water content relative to soils under no cover crops in spring six months after termination when average residue production was 6 Mg ha^{-1} (Blanco-Canqui et al., 2011). The same study found soils under summer cover crops were, on average, 2.2 °C cooler in the upper 30 cm depth in spring compared with soils without cover crops. Thus, soil volumetric water content was negatively correlated with soil temperature, indicating water content in the soil increases with cover crop-induced decrease in soil temperature.
- Cover crops can improve water holding capacity by increasing soil C concentration, particularly in the long term, although evidence of a large contribution of increased soil C concentration to available water appears to be limited (Irmak et al., 2018). Minasny and McBratney (2018) reported that a 1% mass increase in soil organic C concentration can only increase available water capacity by 1.16 mm water per 100 mm of soil. This suggests that even large increases in soil C can have small effects on increasing available water. The magnitude of impact of cover crop-induced increase in soil C on available water can depend on soil-particle size distribution. An increase in soil C concentration can increase available water content more in coarse-textured than in fine-textured soils. Clay particles have similar properties (e.g., size, charge) to soil C and can thus be slower to respond to changes in soil C concentration after cover crop

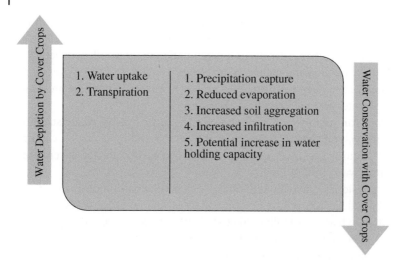

Figure 9.3 Positive, neutral, and negative impacts of cover crops on soil water.

introduction. Fine-textured soils may need larger amounts of soil C than coarse-textured soils to have similar effects.

One of the primary mechanisms by which cover crops maintain soil water, especially in the short term, can be simply by covering or shading the soil with their canopy and residues because cover crop-induced changes in soil properties that affect soil water might not be measurable in the short term. Indeed, changes in soil structural properties and soil C concentration are often observed 5 or 10 years after cover crop introduction (Blanco-Canqui & Ruis, 2020). Well-established cover crops can immediately cover the soil, alter near-surface temperature, minimize evaporation, reduce runoff risks, increasing water capture without significantly changing soil structural properties. In sum, cover crops can have negative, positive, and neutral impacts on soil water dynamics and storage (Figure 9.3).

9.5 Water Management

The impacts of cover crops on soil water are a function of the amount of cover crop biomass produced, tillage system used, timing of cover crop termination, and soil texture, among others. An interaction among these factors affects the extent to which cover crops alter soil water content. For example, an increase in cover crop biomass with late-terminated cover crops can reduce soil water content and thus reduce subsequent crop yields in water-limited regions.

9.5.1 Biomass Production

Cover crop effects on soil water content directly depend on the amount of cover crop biomass production. Soil water content is inversely and significantly correlated with the amount of cover crop biomass produced. In general, the greater the biomass production, the greater the water use by cover crops. Estimates from water-limited regions indicate that the amount of soil water decreases by 0.58 cm for every 1 Mg ha^{-1} of cover crop biomass produced (Blanco-Canqui et al., 2022). High-biomass producing cover crops could deplete water more than low-biomass producing cover crops. For example, a three-year study across four sites in the central Great Plains with precipitation amount ranging from 350 to 917 mm found cover crops when either interseeded or drilled after harvest produced between 0 and 3.2 Mg ha^{-1} and had no effect on soil water content in the 1.2 m depth (Barker et al., 2018). This study shows that low and variable cover crop biomass production can lead to low water use. Wang et al. (2021) found that if cover crops accumulate more than 5 Mg ha^{-1} of biomass, cover crops can significantly deplete soil water and have negative effects on subsequent crop yields. Thus, the threshold level of biomass that minimizes significant soil water depletion for the next crop should be identified for each management system and precipitation zone in water-limited regions.

While it is commonly considered that increased cover crop biomass production will lead to greater transpiration and thus to lower soil water content, a review by Meyer et al. (2019) argued that the relationship between cover crop biomass production and soil water may not be linear. Cover crops use more water to produce large amounts of biomass, but at the same time, the increased aboveground and belowground biomass production can improve soil hydraulic properties more and cover the soil better to reduce evaporation compared with low-biomass producing cover crops. Tradeoffs between water use by cover crops and the soil benefits of high biomass production deserve consideration. High-biomass producing cover crops can use more water than low-biomass producing cover crops, but the former can accumulate more residues in the soil surface after termination, favoring more precipitation capture and soil profile recharge.

Some may suggest cover crop mixes could have greater water use efficiency than monocultures due to potential complementarities among different species within a cover crop mix. However, research indicates cover crop mixes do not use less water than single species. Data show cover crop biomass production can be more critical than differences in the number of species grown. A two-year study at two silt loam sites in the central U.S. Great Plains found soil-profile (1.5–1.8 m) water content between a 10-species cover crop mixture and single-species did not differ (Nielsen et al., 2015). Also, a meta-analysis by Wang et al. (2021) reported that non-legume and legume single species cover crops increased water storage but cover crop mixtures had no effect relative to no cover crops in the upper 30 cm soil depth. Single species cover crops that produce more biomass than mixtures can be more effective at shading soil, reducing evaporation, and enhancing water recharge after termination.

9.5.2 Timing of Cover Crop Termination

Timing of cover crop termination to maximize cover crop biomass production and minimize impacts on soil water is critical. Early termination of cover crops can minimize soil water reduction if sufficient precipitation occurs between cover crop termination and main crop planting. A compromise between early termination and biomass production can minimize water depletion. Compared with late-terminated cover crops, early terminated cover crops can use less water but will also produce less biomass needed to boost soil ecosystem services. For example, the greater the amount of cover crop biomass produced, the greater the potential for erosion control, soil C sequestration, weed suppression, and other services.

Some have suggested that cover crops should be terminated approximately three weeks before main crop planting in water-limited regions to allow sufficient water recharge. Wang et al. (2021) concluded that cover crops should be terminated approximately 20 days before crop planting to allow water recharge as terminating earlier can lead to increased evaporation due to limited residue cover, whereas terminating later may not allow enough time for water recharge. In the western U.S. Corn Belt, late-terminated cover crops reduced soil water content by 37% at corn planting in one of three years when rainfall at corn establishment was lower (60 mm) than in other years (80–120 mm) and reduced corn yields compared to no cover crops in no-till continuous corn (Ruis et al., 2017). This finding indicates that reduced soil water content with late-termination of cover crops can reduce subsequent crop yields when precipitation at crop establishment is lower than normal.

Growing cover crops longer can reduce soil water more than early-terminated cover crops, but, after termination, it can conserve water well into the main crop growing season, potentially increasing crop yields (Burke et al., 2021; Kaspar & Singer, 2011). Timing of termination and the critical importance of precipitation between termination and main crop planting cannot be overemphasized to reduce the potential adverse impacts of cover crops on subsequent crop yields in water-limited regions. In other words, flexible cover crop termination dates, based on the prevailing weather conditions, can be important to manage soil water.

9.5.3 Tillage System

Reducing soil disturbance and mulching the soil surface with cover crop residues such as under no-till management can be strategies to conserve water after cover crop termination in water-limited regions. Cover crops can return significant amounts of residue when biomass production is high, which can be valuable for reducing evaporation when residues are removed in no-till systems. If cover crops are incorporated into the soil as green manure, the benefits of cover crop residues for soil surface protection and reduction in evaporation can be erased or significantly diminished. In cases where no-till adoption is not an option, use of reduced till including strip till, mulch till, and others that minimize soil disturbance and leave most cover crop residues on the soil surface can be a better option than resorting to intensive tillage systems.

Cover crop introduction into no-till systems can fulfill two key principles of no-till management, which are increased crop residue input and intensification of cropping systems. One, no-till technology works when sufficient amount of crop residues is present. However, some no-till cropping systems produce limited amounts of residues in water-limited environments. Thus, the addition of cover crops to low-biomass producing no-till cropping systems can augment the total amount of crop residue produced and contribute to reduced evaporation and increased water storage. Two, the addition of cover crops during fallow periods can diversify the traditional crop-fallow systems and contribute to the goals of agricultural intensification.

The purpose of fallowing is to store water for the next crop, but research shows precipitation storage efficiency during fallow period can be only approximately 30%. Planting cover crops in the fallow period can use some of the 70% of water lost during this period. It is often argued that absence of complex or diversified crop rotations is often a reason for no-till failure. Thus, the addition of cover crops could be a strategy that contributes to no-till diversification and success. However, it is critical to stress that success with cover crops in water-limited regions can only be possible when precipitation is sufficient or near normal. Irrigation is an option to grow cover crops in low precipitation regions (Burke et al., 2021). However, how this practice affects the total water balance in agricultural systems needs consideration.

In high precipitation regions, managing excess water with no-till combined with cover crops can be an alternative to use of tillage. No-till soils are often wetter and cooler than tilled soils. In France, Alletto et al. (2022) reported actual evapotranspiration under faba bean cover crop and vetch–oat cover crop mix was 21–37% greater than under no cover crops. Delaying cover crop termination can be key to use extra water and contribute to improved drainage in wet periods.

9.5.4 Soil Texture

Cover crop impacts on soil water use can also depend on soil texture. Cover crops may need to be managed differently in coarse-textured soils than in fine-textured soils. Because coarse-textured soils have lower ability to hold water than fine-textured soils, cover crops can extract most of the water from coarse-textured soils. Drainage and evaporation rates are also larger in coarse-textured soils relative to fine-textured soils. Timing of cover crop termination and precipitation will be more critical in coarse-textured than in fine-textured soils. It is clear that cover crop implications on soil water are complex as cover crops interact with climate, cover crop management, and soil texture, among others. Overall, cover crops do not reduce water for the next crop in all cases in water-limited environments. Cover crops use water in all cases, but they can also counterbalance such water use under proper management and favorable weather conditions.

9.6 Summary

Impacts of cover crops on soil water are mixed and are threefold. Cover crops have positive, neutral, and negative effects on soil water. Cover crops transpire during their growth and thus reduce soil water in all precipitation zones. Such reduction in soil water with cover crops can be detrimental in water-limited regions, but it can be beneficial in high precipitation regions to improve drainage and reduce excess water. In water-limited regions, cover crops can reduce water for the subsequent crops; however, residue mulch left after cover crop termination can shade the soil, reduce evaporation, and conserve water. Also, cover crop residue mulch can improve soil properties (reduced surface sealing, and increased macroporosity and water infiltration), which can potentially cancel out the negative effects on water use.

Cover crop impacts on soil water depend on cover crop biomass production, cover crop termination timing, timing and amount of precipitation, and potential improvement in soil hydraulic properties. Timely precipitation between cover crop termination and main crop planting can allow soil profile recharge. The longer the cover crop grows, the greater the amount of water used. Identifying optimum termination dates for each climatic region can help to manage cover crop impacts on soil water.

Cover crops can reduce water drainage and groundwater recharge in some cases, but they may or may not reduce soil water content at crop planting if significant precipitation occurs between termination and main crop planting. No-till management can conserve soil water after cover crop termination relative to tilled systems. Overall, cover crops can affect soil water balance, which requires examination of cover crop effects and development of cover crop management strategies on a site-specific basis.

References

Alletto, L., Cassigneul, A. A., Duchalais, S., Giuliano, J., Brechemier, J., & Justes, E. (2022). Cover crops maintain or improve agronomic performances of maize monoculture during the transition period from conventional to no-tillage. *Field Crops Research, 283*, 108540.

Barker, J. B., Heeren, D. M., Koehler-Cole, K., Shapiro, C. A., Blanco-Canqui, H., Elmore, R. W., Proctor, C. A., Irmak, S., Francis, C. A., Shaver, T. M., & Mohammed, A. T. (2018). Cover crops have negligible impact on soil water in Nebraska maize-soybean rotation. *Agronomy Journal, 110*, 1718–1730.

Basche, A., Kaspar, T., Archontoulis, S., Jaynes, D. B., Sauer, T. J., Parkin, T., & Miguez, F. (2016). Soil water improvements with the long term use of a winter rye cover crop. *Agricultural Water Management, 172*, 40–50.

Blanco-Canqui, H., Mikha, M. M., Presley, D. R., & Claassen, M. M. (2011). Addition of cover crops enhances no-till potential for improving soil physical properties. *Soil Science Society of America Journal, 75*, 1471–1482.

Blanco-Canqui, H., & Ruis, S. (2020). Cover crops and soil physical properties. *Soil Science Society of America Journal, 1527–1576.*

Blanco-Canqui, H., Ruis, S., Holman, H., Creech, C., & Obour, A. (2022). Can cover crops improve soil ecosystem services in water-limited environments? A review. *Soil Science Society of America Journal, 86*, 1–18.

Burke, J. A., Lewis, K. L., Ritchie, G. L., DeLaune, P. B., Keeling, J. W., Acosta-Martínez, V., Moore, J. M., & McLendon, T. (2021). Net positive soil water content following cover crops with no tillage in irrigated semi-arid cotton production. *Soil and Tillage Research, 208*, 1–8.

Gabriel, J. L., García-González, I., Quemada, M., Martin-Lammerding, D., Alonso-Ayuso, M., & Hontoria, C. (2021). Cover crops reduce soil resistance to penetration by preserving soil surface water content. *Geoderma, 386*, 114911.

Grunwald, D., Stracke, A., & Koch, H. J. (2022). Cover crop effects on soil structure and early sugar beet growth. *Soil Use and Management.* (in press). https://doi.org/10.1111/sum.12800

Holman, J. D., Arnet, K., Dille, J., Maxwell, S., Obour, A., Roberts, T., Roozeboom, K., & Schlegel, A. (2018). Can cover or forage crops replace fallow in the semiarid Central Great Plains? *Crop Science, 58*, 932–944.

Irmak, S., Mohammed, A. T., Sharma, V., & Djaman, K. (2018). Impacts of cover crops on soil physical properties: Field capacity, permanent wilting point, soil-water holding capacity, bulk density, hydraulic conductivity, and infiltration. *Transactions of the ASABE, 61*, 1307–1321.

Kahimba, F. C., Sri Ranjan, R., Froese, J., Entz, M., & Nason, R. (2008). Cover crop effects on infiltration, soil temperature and soil moisture distribution in the Canadian prairies. *Applied Engineering in Agriculture, 24*, 321–333.

Kaspar, T. C., & Singer, J. W. (2011). The use of cover crops to manage soil. In J. L. Hatfield & T. J. Sauer (Eds.), *Soil management: Building a stable base for agriculture*, Am. Soc. Agron. & Soil Sci. Soc. Am. (pp. 321–337). Wiley.

Kelly, C., Schipanski, M. E., Tucker, A., Trujillo, W., Holman, J. D., Obour, A. K., Johnson, S. K., Brummer, J. E., Haag, L., & Fonte, S. J. (2021). Dryland cover crop soil health benefits are maintained with grazing in the U.S. high and central plains. *Agriculture, Ecosystems and Environment, 313*, 107358.

Korucu, T., Shipitalo, M. J., & Kaspar, T. C. (2018). Rye cover crop increases earthworm populations and reduces losses of broadcast, fall-applied, fertilizers in surface runoff. *Soil and Tillage Research, 180*, 99–106.

Meyer, N., Bergez, J.-E., Constantin, J., & Justes, E. (2019). Cover crops reduce water drainage in temperate climates: A meta-analysis. *Agronomy for Sustainable Development, 39*. http://dx.doi.org/10.1007/s13593-018-0546-y

Minasny, B., & McBratney, A. B. (2018). Limited effect of organic matter on soil available water capacity. *European Journal of Soil Science, 69*, 9–47.

Nielsen, C. C., Lyon, D. J., Hergert, G. W., Higgins, R. K., Calderón, F. J., & Vigil, M. F. (2015). Cover crop mixtures do not use water differently than single-species plantings. *Agronomy Journal, 107*, 1025–1038.

Payero, J., Marshall, M., Davis, R., Bible, J., & Nemire, N. (2021). Effect of rye and mix cover crops on soil water and cotton yield in a humid environment. *Open Journal of Soil Science, 11*, 271–284.

Rankoth, L. M., Udawatta, R. P., Anderson, S. H., Gantzer, C. J., & Alagele, S. (2021). Cover crop influence on soil water dynamics for a corn–soybean rotation. *Agrosystems Geosciences Environment, e20175*, 1–10.

Rosa, A. T., Creech, C. F., Elmore, R. W., Rudnick, D. R., Lindquist, J. L., Butts, L., Pinho de Faria, I. K., & Werle, R. (2021). Contributions of individual cover crop species to rainfed maize production in semi-arid cropping systems. *Field Crops Research*. http://dx.doi.org/10.1016/j.fcr.2021.108245

Ruis, S. J., Blanco-Canqui, H., Jasa, P. J., Ferguson, R. B., & Slater, G. (2017). Can cover crop use allow increased levels of corn residue removal for biofuel in irrigated and rainfed systems? *Bioenergy Research, 10*, 992–1004.

Unger, P. W., & Vigil, M. F. (1998). Cover crop effects on soil water relationships. *Journal of Soil and Water Conservation, 53*, 200–207.

Villarreal, R., Lozano, L. A., Melani, E. M., Polich, N. G., Salazar, M. P., Bellora, G. L., & Soracco, C. G. (2021). First-year cover crop effects on the physical and hydraulic properties of the surface layer in a loamy soil. *Soil and Tillage Research, 213*, 105141.

Vujić, S., Krstić, D., Mačkić, K., Čabilovski, R., Radanović, Z., Zhan, A., & Ćupina, B. (2021). Effect of winter cover crops on water soil storage, total forage production, and quality of silage corn. *European Journal of Agronomy, 130*, 126366.

Wang, J., Zhang, S., Sainju, U. M., Ghimire, R., & Zhao, F. (2021). A meta-analysis on cover crop impact on soil water storage, succeeding crop yield, and water-use efficiency. *Agricultural Water Management, 256*, 107085.

Zhang, Z., Yan, L., Wang, Y., Ruan, R., Xiong, P., & Peng, X. (2022). Bio-tillage improves soil physical properties and maize growth in a compacted Vertisol by cover crops. *Soil Science Society of America Journal, 86*, 324–337.

10

Weed Management

10.1 Cover Crops and Weeds

Cover crops have long been recognized as a potential biological strategy to manage weeds in agricultural lands. Weed control is perhaps one of the major reasons for cover crop adoption after soil erosion and soil fertility management goals. Impacts of cover crops on soil properties are often slow to develop, but well-established cover crops can rapidly contribute to weed control. The weed suppression benefit of cover crops can be essential for the delivery of numerous soil services from agroecosystems including crop production, water conservation, soil fertility, and others.

Management of weeds, particularly herbicide-resistant weeds, is one of the major growing challenges of no-till management (Osipitan et al., 2018; Kumar et al., 2020). Weeds have been traditionally controlled with herbicides especially in large-scale farming. However, some weed species such as kochia have developed resistance to continued use of herbicides (Petrosino et al., 2015). Even newly-developed herbicides with contrasting modes of action do not appear to be fully effective against herbicide-resistant weeds. The number of herbicide-resistant weeds is increasing whereas the number of new herbicides or weed treatment options is decreasing (Kumar et al., 2020).

Development of herbicide-resistant weeds is particularly a problem in croplands with limited or no soil disturbance such as no-till systems. Growing concerns over herbicide-resistant weeds are leading to the consideration of one-time or strategic tillage once in 5 or 10 years of long-term no-till fields (Blanco-Canqui & Wortmann, 2020). Strategic deep tillage could be an option to bury weed seeds and reduce weed germination and growth in no-till cropping systems. Published literature indicates that one-time tillage does not generally reduce the soil

Cover Crops and Soil Ecosystem Services, First Edition. Humberto Blanco.
© 2023 American Society of Agronomy, Inc. / Crop Science Society of America, Inc.
/ Soil Science Society of America, Inc. Published 2023 by John Wiley & Sons, Inc.

ecosystem services provided by long-term no-till soils but could increase risks of soil erosion immediately after tillage before vegetation is established (Blanco-Canqui & Wortmann, 2020). As result, other strategies, particularly biological strategies, are sought as potential alternatives to tillage. This chapter discusses the effectiveness of cover crops to suppress weeds and the factors that can influence such effectiveness.

10.2 Weed Suppression

Literature reviews indicate that cover crops can be highly effective at reducing weed biomass (Figure 10.1). In general, cover crops can reduce weed biomass by 90–100% (Osipitan et al., 2018; Blanco-Canqui et al., 2020). Thus, cover crops can be an effective management practice for suppressing weeds. Note that cover crop effects on other ecosystem services such as soil properties and C sequestration and dynamics are often mixed and slow to develop, but cover crops can immediately suppress weeds when successfully established. Cover crops can be particularly useful early in the crop season when weed competition with crops is the greatest (Osipitan et al., 2018; Kumar et al., 2020). The level of weed suppression provided

Figure 10.1 Winter rye cover crop can be effective at suppressing weeds (left) compared to no cover crop (right) Photo by H. Blanco.

by cover crops can be similar to that provided by herbicides particularly for early season weeds (Osipitan et al., 2018). Published data indicate that cover crops can reduce weed proliferation by physical, chemical, and biological mechanisms. Specifically, they can directly and indirectly control weeds by:

1) Competing with weeds for space, water, essential nutrients, and light.
2) Shading soil surface and reducing weed germination and growth.
3) Smothering weeds via physical mechanisms.
4) Acting as a physical barrier to weed growth.
5) Releasing toxic compounds (e.g., allelochemicals) that suppress weed growth.

It is important to discuss, however, that cover crops can be most effective at suppressing weeds when cover crops are growing or right after cover crop termination when residue amount is high. The weed-suppressing ability of cover crops diminishes with time after termination due to rapid residue decomposition (Osipitan et al., 2018). Thus, cover crops can be more effective at suppressing early season weeds than late season weeds during the crop growing season. Combining cover crops for early season weed suppression with herbicide application for mid or late season weed suppression can be a more viable approach to manage weeds compared with either using herbicide or cover crop alone.

While cover crops may not completely eliminate the need for herbicides, the high effectiveness of cover crops for controlling weeds during critical periods of crop establishment indicates that cover crops can significantly reduce the use of herbicides. Herbicides could be required for only a portion of the crop growing season. Also, mechanical termination of cover crops can be an alternative to chemical termination to reduce use of herbicides. Mowing, rolling with a crimper, green manuring, and undercutting or slicing cover crops underneath the soil are some of methods for mechanical termination of cover crops.

The potential of cover crops to manage weeds is promising, but the magnitude of weed control with cover crops can depend on a number of factors including cover crop biomass production, cover crop species, planting and termination timing, tillage and cropping systems, and climate. How these factors affect the ability of cover crops to suppress weeds deserves discussion based on published information. For example, cover crops established before weeds emerge can be more effective at controlling weeds than those established after weeds have emerged. Interseeding cover crops with main crops can allow weed suppression as crops develop, but interseeded cover crops could compete with main crops for resources, which can potentially reduce crop yields.

Some strategies to enhance cover crop potential to suppress weeds include boosting biomass production, planting high-biomass producing species, establishing early before weed emergence, and combining cover crops with reduced tillage and herbicide application (Figure 10.2). In sum, experimental data indicate that cover

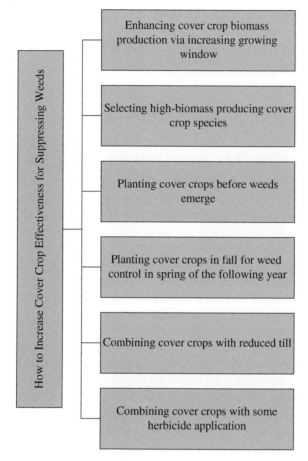

Figure 10.2 Some potential strategies that can enhance the effectiveness of cover crops for managing weeds within integrated weed management programs.

crops can be an effective tool to manage weeds, particularly early season weeds, and can be an essential component of integrated weed management programs.

10.3 Managing Weeds

Among the factors that affect the ability of cover crops to suppress weeds include amount of cover crop biomass produced, percent surface cover, cover crop mixes vs. single species, tillage system, and climate. For instance, the greater the amount of biomass produced and the larger the percentage of surface covered with cover crops, the greater the weed suppression potential of cover crops. However, differences in precipitation input can determine the amount of cover crop biomass produced, surface cover, and thus the ability of cover crops to control weeds.

10.3.1 Biomass Production and Surface Cover

One of the key factors that affects the effectiveness of cover crops for controlling weeds is the amount of cover crop biomass produced (Figure 10.2). The weed suppression effectiveness of cover crops increases as the amount of cover crop biomass increases. Thus, high-biomass producing cover crop species are most effective at reducing weed biomass. In a review of 53 studies, Osipitan et al. (2019) reported that weed biomass and density were significantly and inversely correlated with cover crop biomass production. Cover crop residues left on the soil surface after termination of high-biomass producing cover crops can also persist and thus suppress weeds well into the main crop growing season. Growing cover crops can provide weed suppression benefits before and at planting main crops, while cover crop residues when abundant can allow weed suppression after crop planting till crops are well established. In other words, abundant production of cover crop residues can extend cover crops benefits well into the main crop season.

There may be a threshold level of biomass required to suppress weeds. Cover crops may have minimal effects on reducing weed biomass if they produce less than $1\,Mg\,ha^{-1}$ (Kumar et al., 2020). This minimum level of biomass production needed for weed suppression is similar to that needed for improving other soil ecosystem services. Approximately $4–5\,Mg\,ha^{-1}$ of cover crop biomass may be needed for optimum weed suppression (Kumar et al., 2020; Osipitan et al., 2019). The threshold level of cover crop biomass production should be established for different weed species as the amount of biomass needed for suppression may vary with weed species.

Increasing cover crop biomass production for weed suppression depends on cover crop growing window. As discussed earlier, planting cover crops earlier and terminating them late can produce more biomass than late-planted and early-terminated cover crops. Delaying cover crop termination even for one week can significantly increase biomass production and thus suppress weeds more effectively than early-terminated cover crops. Research shows that delaying cover crop termination by approximately 10 days can double biomass production (Ruis et al., 2017). However, because growing cover crops longer can use more water than early-terminated cover crops, timing of planting and termination is important in water-limited regions. Other strategies to boost cover crop biomass production and thus weed suppression include increasing seeding rates, and irrigating and fertilizing cover crops although the economic implications of the practices warrant further analysis.

More research focuses on the amount of cover crop biomass produced and less on the soil surface cover provided by cover crops before and after termination. The percentage of soil surface covered by cover crops is as important if not more important than biomass production. Surface cover determines shading, light blocking, physical barrier, and other processes that limit weed growth (Figure 10.1). For example, light quantity and quality can decrease with an increase in soil surface coverage. Cover crops with broad leaves can cover soil better than those with narrow leaves such as winter rye, but the amount of biomass produced could overcome the differences in leaf architecture.

Weed suppression increases as the percentage of surface covered by residues increases. Dense and uniform surface cover by cover crops is needed to fully control weed growth. Depending on row spacing, cover crops drilled in rows often maximize surface coverage along the rows and may not uniformly cover the inter-row positions. Seeding methods such as broadcast seeding that distribute cover crop seeds across the whole field may allow a more uniform soil coverage than drilled in rows although this seeding method does not always produce more biomass than drilled cover crops. Similarly, flat cover crop residues can provide better coverage than standing cover crops. Also, cover crops that keep the soil covered for extended periods of time after termination can provide better weed suppression.

10.3.2 Cover Crop Species and Mixes

Literature indicates that grass cover crops can compete better with weeds than legume and brassica cover crops because grasses decompose slower and thus their weed-suppressing effects can last longer. Also, grass cover crops can release limited amounts of nutrients upon decomposition and thus reduce weed growth relative to legume cover crop residues (Osipitan et al., 2019). Among the potential grass cover crop species that can suppress weeds include sorghum-sudangrass, winter rye, triticale, wheat, and oat. As an example, under corn and soybean systems in the U.S. Midwest, winter rye is one of the most common species used to suppress weeds because of its low cost, high availability, easy establishment, and winter hardiness. Winter rye can also produce large amounts of biomass in spring, especially when termination date is delayed till main crop planting. Winter rye is a dominant species when included in cover crop mixes even when seeding rate is lowered. It is well known that an increase in seeding rate commonly increases biomass production, but the extent of this increase is related to cover crop species (Werle et al., 2017). Grass cover crops can produce more biomass than broadleaf cover crops under the same rate of seeding. Thus, a higher seeding rate can be required for broadleaves to produce the same amount of biomass as grasses.

Winter annual weeds commonly emerge in the fall and grow until spring or early summer. Thus, growing winter grass cover crops during this period can suppress the winter annual weeds and reduce the use of herbicides in fall and spring. Matching cover crop growth with the growth cycle of annual weeds can be an effective strategy to control weeds (Werle et al., 2017). Cover crops planted in late summer or early fall and terminated in late spring can suppress weeds in fall, winter, and spring and reduce the need for herbicide use. However, cover crops planted in late summer or early fall and terminated in early spring can have limited effectiveness to suppress weeds in spring. It is also important to note that often herbicides are used to terminate cover crops, which may not thus completely eliminate the need for herbicides in a cropping system.

Overall, broadleaf cover crops may not be as effective as grass cover crops to control weeds (MacLaren et al., 2019; Osipitan et al., 2019). When broadleaf cover crop species are preferred, hairy vetch and clover can be options. Hairy vetch is one of the most winter hardy legumes, which can suppress weeds in fall, winter, and spring. While cover crop species can reduce weed growth in this order: Grasses > Legumes = Brassica, all cover crops can suppress weeds if they produce significant amounts of biomass (Kumar et al., 2020; Osipitan et al., 2019). Cover crop biomass production and cover crop residue persistence on the soil surface can be more important than cover crop diversity for controlling weeds. Overall, grass cover crops could suppress weeds more than other cover crops, but the total amount of biomass produced is more important for weed management than cover crops species.

Because cover crop mixtures produce similar amounts of biomass to single species (Holman et al., 2018), both mixtures and single species can be equally effective at suppressing weeds. However, planting high-producing single species cover crops can be less expensive alternative than cover crop mixtures to manage weeds. Research shows cover crop diversity is less important than biomass production for suppressing weeds (MacLaren et al., 2019). The idea that simply mixing cover crops will lead to multiple services including effective weed suppression is not supported by experimental data.

Selection of cover crops that complement each other and produce the highest amount of biomass when mixed can be key to control weeds. Because cover crop biomass production is the best predictor of weed suppression, high-biomass producing single species cover crops can be more effective at suppressing weeds than a mixture that produces low amounts of biomass. However, if surface cover at different times is needed for weed control, planting a mix of different species may be favored. As mentioned earlier, some cover crop species (e.g., radish) grow rapidly in fall, while others grow rapidly in spring (e.g., winter rye; Wallace et al., 2021). Thus, planting a mix in fall can potentially provide surface cover and suppress weeds at different times till main crops are planted the following year.

10.3.3 Tillage System

Influence of tillage on cover crop ability to suppress weeds can be complex. Because tillage itself can suppress weeds due to soil disturbance, weed suppression with the incorporation of cover crops does not imply that burying cover crop residues is more effective for weed control than leaving residues on the soil surface. Cover crops are often combined with tillage to manage weeds in tilled environments (Kumar et al., 2020). Traditionally, tillage was primarily used to manage weeds. However, the advent of herbicides such as glyphosate [N-(phosphonomethyl) glycine] in the mid 1990s favored increased adoption of no-till technology without the need of tillage for weed control.

Lack of tillage in fall or spring has allowed proliferation of weeds, such as herbicide resistant weeds. Cover crops may need to be combined with some soil disturbance to fully control weeds. In a review, Osipitan et al. (2019) concluded cover crops suppressed weeds more in reduced till than in no-till systems. Completely eliminating tillage such as in no-till and solely relying on herbicides or cover crops for weed control may not fully control herbicide resistant weeds. Under no-till systems, cover crop residue mulch may not be sufficient to combat or eliminate some of the weeds that are highly resistant to herbicides. As discussed earlier, in some cases, a well-planned strategic tillage in long-term no-till systems may be needed for deep placement of weed seeds and interruption of growth cycle of weeds.

10.3.4 Climate

Cover crops can be effective at reducing weed biomass when biomass production is sufficient. However, biomass production is a function of climate in addition to cover crop management. Cover crops can be more successfully grown in high precipitation regions for weed suppression than in low precipitation regions due to more water availability in high precipitation regions. Cover crop establishment can be a challenge in water-limited regions, warranting the need for flexibility in cover crop use. Cover crops may need to be planted only in wet years or when precipitation is near normal.

Studies from semiarid regions where cover crops have been used to replace fallow in crop-fallow systems indicate that cover crops offer potential for suppressing weeds. A review with specific focus on semiarid regions reported that cover crops can reduce weed biomass by approximately 90% in these regions (Blanco-Canqui et al., 2020). Also, at two sites in the U.S. central Great Plains with an average annual precipitation of approximately 600 mm, winter rye planted after corn silage harvest in irrigated continuous corn silage systems under conservation tillage reduced weed biomass by 91–95% and weed density by 91% compared to fallow systems without cover crops (Werle et al., 2017). At both sites, cover crop biomass production ranged from 3.76 to 4.08 Mg ha^{-1}, which is close to the optimum amount of cover crop biomass needed to suppress weeds and improve other soil ecosystem services.

Depending on management (e.g., planting and termination timing) and cropping system, cover crops such as winter rye can produce significant amounts of biomass and be highly effective for suppressing weeds in water-limited regions. Werle et al. (2017) found that winter rye effectively suppressed horseweed, a herbicide-resistant weed, in water-limited regions when average winter rye biomass production was 8.6 ± 3.3 Mg ha^{-1}. Also, while grass cover crops can produce significant amounts of biomass under long growing window such as from late summer to late spring of the following year, termination before flowering is recommended to reduce negative impacts on the subsequent crop yield in water-limited regions. In water-limited regions, irrigation of cover crops may be needed to establish and produce sufficient amount of biomass for controlling weeds.

10.4 Summary

Cover crops are an effective biological strategy to manage weeds, particularly for early crop season weeds. Introduction of cover crops can suppress weeds by 90–100% and can potentially reduce the total amount of herbicides needed for the crops. Cover crops can be used to suppress early crop season weeds, while herbicides can be used to control weeds after crop establishment. Cover crops suppress weeds by physically shading and reducing space for weed growth, biologically competing with weeds for resources, and chemically emitting allelopathic compounds that inhibit weed proliferation.

Factors affecting cover crop effectiveness at suppressing weeds can be in this order: Biomass Amount = Surface Cover > Cover Crop Species > Tillage Systems > Cover Crop Mixes or Single Species, indicating that biomass production is the top predictor of weed suppression. Practices that increase cover crop biomass production can increase cover crop effectiveness for weed suppression. Approximately 4 Mg ha^{-1} of cover crop biomass is needed for optimum weed suppression.

Grass cover crops can be more effective at suppressing weeds than legume cover crops due to slower residue decomposition and quicker establishment of grass cover crops. Cover crop effects may not differ between mixes and single cover crop species. Establishing cover crops early before weed emergence, and combining cover crops with reduced tillage and herbicide application can be strategies to enhance cover crop potential for weed suppression. Overall, cover crops can be an essential tool within the portfolio of integrated weed management practices to manage weeds including herbicide-resistant weeds.

References

Blanco-Canqui, H., Ruis, S., Holman, H., Creech, C., & Obour, A. (2020). Can cover crops improve soil ecosystem services in water-limited environments? A review. *Soil Science Society of America Journal, 86*, 1–18.

Blanco-Canqui, H., & Wortmann, C. W. (2020). Does occasional tillage undo the ecosystem services gained with no-till? A review. *Soil and Tillage Research, 198*, 104534.

Holman, J. D., Arnet, K., Dille, J., Maxwell, S., Obour, A., Roberts, T., Roozeboom, K., & Schlegel, A. (2018). Can cover or forage crops replace fallow in the semiarid Central Great Plains? *Crop Science, 58*, 932–944.

Kumar, A. V., Obour, A., Jha, P., Manuchehri, M. R., Dille, J. A., Holman, J., & Stahlman, P. W. (2020). Integrating cover crops for weed management in the semi-arid U.S. Great Plains: Opportunities and challenges. *Weed Science, 68*, 311–323.

MacLaren, C., Swanepoel, P., Bennett, J., Wright, J., & Dehnen-Schmutz, K. (2019). Cover crop biomass production is more important than diversity for weed suppression. *Crop Science, 59*, 733–748.

Osipitan, O. A., Dille, A., Assefa, Y., Radicetti, E., Ayeni, A., & Knezevic, S. Z. (2019). Impact of cover crop management on level of weed suppression: A meta-analysis. *Crop Science, 59*, 833–842.

Osipitan, O. A., Dille, J. A., Assefa, Y., & Knezevic, S. Z. (2018). Cover crop for early season weed suppression in crops: Systematic review and meta-analysis. *Agronomy Journal, 110*, 2211–2221.

Petrosino, J. S., Dille, J. A., Holman, J. D., & Roozeboom, K. L. (2015). Kochia suppression with cover crops in southwestern Kansas. *Crop, Forage & Turfgrass Management, 1*, 1–8.

Ruis, S. J., Blanco-Canqui, H., Jasa, P. J., Ferguson, R. B., & Slater, G. (2017). Can cover crop use allow increased levels of corn residue removal for biofuel in irrigated and rainfed systems? *Bioenergy Research, 10*, 992–1004.

Wallace, J., Isbell, S., Hoover, R., Barbercheck, M., Kaye, J., & Curran, W. (2021). Drill and broadcast establishment methods influence interseeded cover crop performance in organic corn. *Renewable Agriculture and Food Systems, 36*, 77–85.

Werle, R., Burr, C., & Blanco-Canqui, H. (2017). Cereal rye cover crop suppresses winter annual weeds. *Canadian Journal of Plant Science*. https://doi.org/10.1139/CJPS-2017-0267

11

Soil Fertility

11.1 Soil Fertility Management

It is well recognized that intensive cultivation has reduced soil fertility due to increased soil erosion, loss of organic matter, and deterioration of soil physical, chemical, and biological processes and properties. Maintaining or improving soil fertility is an increasing global challenge. Inorganic fertilization has allowed increased crop production but not without unintended environmental consequences such as increased risks for water pollution and N_2O emissions. Incorporating practices that not only provide essential nutrients and reduce the sole reliance on inorganic fertilizers but also improve or restore the ability of the soil to adsorb and store nutrients is imperative. Biological practices that improve the chemical resilience of soils for adequate retention and delivery of essential nutrients would also improve the overall resilience of the soil. One such practice can be the use of cover crops.

Cover crops have long been used to improve soil fertility. Ancient civilizations approximately 3000 years ago used cover crops as green manure to improve soil fertility (Groff, 2015). In the U.S., in the early 1900s, Bennett (1939) discussed the benefits of cover crops for halting soil degradation and improving soil fertility. Indeed, before the advent of inorganic fertilizers or Haber-Bosch N, biological N fixation by legume cover crops was probably one of the main sources of N for growing crops. Cover crops were often incorporated into the soil as green manure to accelerate release of nutrients for the subsequent crops.

The reemergence of cover crops has the potential to improve soil fertility, particularly in degraded lands. Cover crops can benefit soil fertility by a myriad of mechanisms including (a) scavenging and taking up nutrients, (b) fixing N from the atmosphere, (c) reducing nutrient leaching, (d) improving soil physical,

Cover Crops and Soil Ecosystem Services, First Edition. Humberto Blanco.
© 2023 American Society of Agronomy, Inc. / Crop Science Society of America, Inc.
/ Soil Science Society of America, Inc. Published 2023 by John Wiley & Sons, Inc.

chemical, and biological properties, (e) reducing soil gas (e.g., N_2O) emissions, and (f) improving overall nutrient cycling. Following termination, decomposition of aboveground and belowground cover crop biomass can release nutrients, organic acids, and other substances that alter nutrient concentration in the soil and ion exchange capacity. This chapter discusses how cover crops affect relevant soil fertility properties including organic matter, nutrients, pH, and cation exchange capacity.

11.2 Organic Matter

Soil organic matter is probably the most important soil fertility indicator. It is a reservoir of essential nutrients including N, P, and S. It also mediates changes in soil physical, chemical, and biological properties, thereby affecting overall soil fertility and productivity. An increase in soil organic matter concentration generally improves soil biological activity, ion exchange capacity, soil aggregation, and pore space, moderates soil temperature, and enhances other soil properties with an extent depending on the amount of increase in organic matter concentration. Increasing organic matter with cover crops to 4% or 5% on a soil volume basis is an ideal goal to boost soil fertility and productivity. However, soil organic matter concentration hardly exceeds 1% in many croplands especially in degraded, water-limited, and tropical environments (Obi, 1999; Santos et al., 2022).

The introduction of cover crops generally increases soil organic matter concentration. A summary of case study examples in Table 11.1 shows that cover crops increased soil organic matter by, on average, 1.20-fold. Additionally, several reviews reported that cover crops generally increase soil organic C concentration (Blanco-Canqui et al., 2022; Poeplau & Don, 2015). Soil organic matter contains approximately 58% C. Thus, an increase in soil C concentration with cover crops often indicates an increase in soil organic matter concentration. Initial soil organic matter level can influence the magnitude of cover crop impacts on soil organic matter. Cover crops can increase organic matter more rapidly in low initial than in high initial organic matter soils. Indeed, Table 11.1 shows that cover crops can significantly increase soil organic matter when initial organic matter concentration is below 3% but may have small or mixed effects when initial organic matter is above 3%. Also, incorporating cover crops as green manure can more rapidly increase organic matter and improve soil fertility for the subsequent crop due to increased decomposition compared with cover crop residues left on the soil surface although tillage can have adverse effects on other soil properties and processes. In a few cases, similar to soil organic C, cover crops can reduce soil organic matter concentration due to the positive priming effect of fresh organic matter from cover crops (Santos et al., 2022).

Table 11.1 Cover Crop Impacts on Soil Organic Matter and Cation Exchange Capacity Across Different Environments in the Upper 30 cm Depth

Location	Soil	Initial organic matter (%)	Cover crops	Duration (yr)	Soil organic matter (%)	Cation exchange capacity
[a]Prince Edward Island, Canada	Sandy loam	2.3	Timothy, buckwheat, brown mustard, and mix (winter rye, hairy vetch, and crimson clover)	1	Increased to 3.1	na
[b]Sao Paulo, Brazil	Sandy	Low	Single species and mixes	3	Increased	Increased
[c]Missouri, U.S.	Loam (eroded)	2.7	Mix of winter cover crops (rye, barley, triticale, and oat)	6	Increased to 3.4	na
[d]New York, U.S.	Silt loam	2.7	Ryegrass, red clover, crimson clover, and hairy vetch	4	Increased to 2.9	na
[e]Nsukka, Nigeria	Sandy clay loam	1.1	Carpet grass, creeping grass, guinea grass, elephant grass, style, and Kudzu	5	Increased to 1.4	Increased from to 5.5 to 5.90 dS/m
[f]Nebraska, U.S.	Silty clay loam	4.9	Cereal rye, grain sorghum, Austrian winter pea, and soybean	12	Increased with grasses; No change with legumes	No change
[g]Piaui State, Brazil	Sandy clay loam	0.8	Multiple single species and mixes	3	Decreased	Decreased
[h]Foulum, Denmark	Sandy loam	3.1	Radish	5	No effect	na

[a] Khan et al. (2021).
[b] dos Santos Cordeiro et al. (2021).
[c] Rankoth et al. (2019).
[d] Nunes et al. (2018).
[e] Obi (1999).
[f] Blanco-Canqui and Jasa (2019).
[g] Santos et al. (2022).
[h] Abdollahi and Munkholm (2014).
na = not available

11.2.1 Nitrogen

Cover crops can increase N concentration in the soil during the fallow period compared to fields without cover crops by (a) scavenging N, (b) maintaining N levels or reducing N losses, and (c) fixing N from the atmosphere. Legume cover crops can increase soil N concentration by fixing atmospheric N, whereas non-legume cover crops can maintain soil N levels by reducing losses of N from the soil. Thus, cover crops can be especially beneficial to manage N relative to other nutrients. For example, cover crops may not only maintain N levels in the soil but can also accumulate N from the atmosphere under favorable conditions.

11.2.2 Nitrogen Scavenging

Cover crops, particularly non-legume cover crops, scavenge N and immobilize N in their tissues by transforming mobile N (inorganic N) into immobile N (organic N; Thapa et al., 2018). The amount of N scavenged per year will vary depending on cover crop biomass production and cover crops species. Cover crops accumulate nutrients in both aboveground and belowground biomass. For example, Sievers and Cook (2018) found N content for winter rye was $11.5\,g\,kg^{-1}$ in the aboveground biomass and $8.3\,g\,kg^{-1}$ in the root biomass, while, for hairy vetch, it was $41.9\,g\,kg^{-1}$ in the aboveground biomass and $16.5\,g\,kg^{-1}$ in the root biomass. Thus, the amount of N immobilized per year could be $30\,kg\,N\,ha^{-1}$ for winter rye and $97\,kg\,N\,ha^{-1}$ for hairy vetch if cover crops would produce $2\,Mg\,ha^{-1}$ of aboveground biomass production and $1\,Mg\,ha^{-1}$ of aboveground biomass production. Note that belowground cover crop biomass production is approximately 40–50% of the aboveground biomass production (Blanco-Canqui et al., 2020).

11.2.3 Reduction of Nitrogen Losses

Cover crops reduce losses of N and maintain N levels in the soil by reducing nitrate leaching, N_2O emissions, and N runoff. A reduction in nitrate concentration in the soil due to N uptake by cover crops directly reduces N available for runoff, leaching, and emissions. Approximately 50% or more of N applied with fertilizers is lost from croplands due to the low use efficiency of N (Thapa et al., 2018). Leaching of nitrates into groundwater is increasingly a major concern. Cover crops can reduce nitrate leaching by 18–95% in most studies as discussed in an earlier chapter (Blanco-Canqui, 2018). Grass cover crops such as winter rye can reduce leaching more than legume cover crops.

Non-legume cover crops generally reduce N_2O emissions, but legume cover crops can increase N_2O emissions (Basche et al., 2014). Non-legume cover crops can reduce N₂O emissions due to immobilization of N and slow residue decomposition relative to legume cover crops. Incorporating cover crops into soil as green manure can accelerate residue decomposition and increase N concentration but may increase risks of nutrient loss and soil erosion. Because cover crops reduce runoff volume and sediment loss, addition of cover crops can reduce losses of sediment-associated N although the effectiveness of cover crops for reducing losses of dissolved N in runoff is mixed.

11.2.4 Nitrogen Fixation

It is well recognized that legume cover crops capture and fix significant amounts of N from the atmosphere, thereby increasing nitrate concentration in the soil and potentially reducing the amount of N fertilizer needed. In a review, Tonitto et al. (2006) concluded that input of N from legume cover crops ranges from 8 to 350 kg N ha^{-1} with 50% of studies reporting between 50 and 150 kg N ha^{-1} of N input. Also, a long-term study in the eastern U.S. Great Plains under no-till winter wheat–grain sorghum rotation found that legume (hairy vetch, sunn hemp, and late-maturing soybean) cover crops increased soil total N by 270 kg ha^{-1} after 15 years (Blanco-Canqui et al., 2012).

Differences in crop yields between fields with legume cover crops and those fertilized with inorganic N at recommended rates may not be significant when legume cover crops produce between 110 and 180 kg N ha^{-1} (Tonitto et al., 2006), which suggests if legume cover crops produce at least 110 kg N ha^{-1}, use of inorganic fertilization may not be needed. However, when legumes produce <110 kg N ha^{-1}, yields of the subsequent crops can be lower (Tonitto et al., 2006). Legume cover crops can be more beneficial to provide N in systems with low N fertilizer input such as organic farming relative to systems with high N fertilizer input. Also, combining legume cover crops with inorganic fertilization by adjusting for N credits from cover crops is a strategy to reduce inorganic fertilizer use.

Release of N from legume cover crops can vary depending on cover crop species. In a tropical climate in Brazil, velvet bean, pearl millet, dwarf pigeon pea, sunn hemp, showy rattlebox, and jack bean summer cover crops released as much as 137 kg N ha^{-1} 35 days after termination in a two-year study under no-till (Weiler et al., 2019). The same study found sunn hemp released less N than other cover crops, while velvet bean and showy rattlebox released N more slowly than other cover crops (Weiler et al., 2019). Differences in the C to N ratio among cover crop species determine N release rate as discussed later (Sievers & Cook, 2018). Even within the same species, the C to N ratio differs between leaves and stems (Mansoer et al., 1997). Stems have a higher C to N ratio than leaves due to structural components like lignin. Thus, cover crops (e.g., sunn hemp) with greater biomass allocation to stems than to leaves can decompose slower than cover crops with more biomass allocated to leaves (e.g., soybean).

While it is assumed that legume cover crops are effective at fixing atmospheric N, research also shows that such effectiveness can be limited, in some cases. Differences in legume cover crop species, soil water content, soil acidity, and microbial diversity or community structure, among others can affect legume-rhizobium symbiosis and the extent of N fixation. In a subtropical fine sandy loam in Texas, U.S., Kasper et al. (2019) found that crimson clover, hairy vetch, and field pea showed limited nodulation after inoculation with rhizobia. One of the factors that can reduce root nodulation and N fixation is low soil water content during nodulation, which is often the case in water-limited environments (Kasper et al., 2019). Also, high N availability in the soil may limit N fixation by cover crops.

11.3 Phosphorus

Cover crops can be an important source of P and thus reduce the amount of inorganic P fertilizer needed for the subsequent crops. Cover crop residue decomposition not only releases N but also P and other essential nutrients. In a review, Liu et al. (2019) reported that the aboveground biomass component of cover crops can contain as much as 26 kg P ha^{-1} in cool temperate regions. If we assume that cover crop root biomass input is approximately 40% of the aboveground cover crop biomass production (Blanco-Canqui et al., 2020), then the total amount of P present in cover crops can be approximately 36 kg P ha^{-1}.

Cover crops can be most valuable for increased P availability in tropical and degraded soils with low available P (Hallama et al., 2019). Phosphorus-deficient soils can have more sorption sites to retain the cover crop-derived P compared with soils with high P saturation levels. In a humid subtropical climate, introduction of oat, rye, and ryegrass cover crops released between 2 and 16 kg of available P during the main crop season (Varela et al., 2017). Cover crops can also improve P cycling in agroecosystems by enhancing microbial biomass and activity (Kim et al., 2020).

Concentration of P and subsequent P release can differ with cover crop species, aboveground biomass production, root biomass production (e.g., tuber versus fibrous roots), cover crop development stage, and climate (Hallama et al., 2019). In a temperate no-till corn–soybean system with and without winter rye and hairy vetch cover crops, Villamil et al. (2006) found soil available P concentration was 30.4 mg kg^{-1} across corn–winter rye/soybean–winter rye sequence and corn–winter rye/soybean–winter rye + hairy vetch sequence and 36.3 mg kg^{-1} under no cover crops. These results suggest that cover crops can immobilize and thus reduce P losses in the off-season. Cover crop species such as legumes can more rapidly release P than grass cover crops due to increased decomposition. Also, the amount of P released can depend on cover crop species. Brassica cover crops (e.g., roots) have higher P concentration than legume and grass cover crops and can thus release more P than other cover crops upon decomposition. In a review, Liu et al. (2019) found that mean P concentration in cover crops was in this order: Brassicas (4.6 mg kg^{-1}) > Grasses (3.6 mg kg^{-1}) > Legumes (3.5 mg kg^{-1}).

Cover crops can also reduce the need for P fertilizer input by reducing P losses with runoff and sediment. Research shows that high-biomass producing cover crops such as grass cover crops are an effective strategy to reduce runoff and losses of sediment and sediment-associated-P. While N is lost via leaching, emissions, and erosion, the main pathway of P loss from agricultural lands is water erosion especially when P is stratified near the surface such as in no-till soils. However, managing losses of P from agricultural lands to improve P use efficiency and reduce risks of non-point source pollution of water sources is an increasing challenge even under cover crops. Indeed, the potential of cover crops to reduce losses of P in runoff can be inconsistent.

As discussed earlier, while cover crops are effective at reducing sediment loss and sediment-associated P, they may not be effective at reducing concentration of dissolved P in runoff. In the central Great Plains, Carver et al. (2022) found that small grain and brassica cover crop mix reduced sediment loss but increased dissolved reactive P loss in three of four years compared with no cover crops. Also, cover crops can release significant amounts of dissolved P after termination. When P is released during the non-growing season, some of the dissolved P can be rapidly lost in runoff or drainage tiles (Liu et al., 2019). For example, in cool temperate regions with frequent freeze–thaw cycles, frozen cover crop biomass can release significant amounts of P as it thaws, which can be susceptible to losses in snowmelt runoff (Lozier et al., 2017). However, note that the amount of dissolved P lost in runoff can be smaller than that retained in the soil, depending on the amount of runoff (Liu et al., 2019). The magnitude of losses of dissolved P in runoff depends on runoff amount and cover crop management. Losses of dissolved P from cover crops can be a concern in regions with intense rainstorms and in regions with severe freeze–thaw cycles accompanied by snowmelt runoff. Literature suggests that while cover crops can be a source of P for crops and reduce sediment-bound P in runoff, cover crops could induce some losses of dissolved P compared with fields without cover crops.

11.4 Other Nutrients

Cover crops can scavenge other nutrients besides N and P and reduce their losses from the soil. For example, in a temperate sandy soil, radish cover crops increased exchangeable K concentration in the topsoil compared to no cover crops after five years of management (Abdollahi and Munkholm, 2014). The increased exchangeable K with cover crops indicates cover crops can reduce losses of K by reducing leaching and erosion, especially in regions where precipitation amount is greater than evaporation. Similarly, on a temperate soil, Khan et al. (2021) discussed cover crops can return K and Ca to soil upon termination.

As with other soil nutrients, cover crops can be particularly useful to reduce losses of K, Ca, and Mg in low fertility soils. For instance, in a degraded acidic tropical soil, addition of legume cover crops increased the concentration of Ca, Mg, and K and reduced concentration of Al after five years (Obi, 1999). Also, in a low organic matter tropical soil, cover crops increased Ca, Mg, and K concentration after three years relative to no cover crops (Santos et al., 2022). These results show that acidic soils with low concentration of basic cations (Ca, Mg, and K) can benefit from cover crops to improve their fertility. It is important to note the extent of increase in the concentration of K, Ca, Mg, and other nutrients with cover crops can be modest in some cases. Increased biomass production and selection of appropriate cover crop species such as legumes can be important to manage soil fertility.

11.5 Soil pH

Soil pH is a measure of soil acidity and alkalinity and is an important indicator of the availability of essential nutrients and soil productivity. It determines the acid and base buffering capacity of the soil. For example, soils with pH < 5 can be deficient in nutrients (P, K, Ca, and Mg) and have limited biological activity but can have high levels of Al and H, which adversely affect soil fertility and productivity. Biological practices that maintain soil pH between 5 and 8 can be ideal to sustain availability of essential nutrients. The question is: Can cover crops be used to manage soil pH? Experimental data indicate cover crop impacts on soil pH can be inconsistent. Table 11.2 summarizes some recent case studies to illustrate how soil pH responds to cover crop introduction. Cover crops can either increase or have no effect on soil pH. It is interesting to note cover crops do not reduce soil pH, which indicates cover crops do not result in soil acidification.

When cover crops increase soil pH, such increase occurs in acidic soils (pH < 7) based on the case studies (Table 11.2). Soil pH can increase by 0.2–0.8 units. The increase in soil pH has positive implications for the management of soil acidity.

11.5 Soil pH

Cover crops can have a liming effect and could ameliorate soil acidity (Santos et al., 2022). Cover crops could reduce soil acidity by reducing losses of basic cations such as Ca, Mg, K, and Na. However, the level of increase in soil pH with cover crops is generally small, and any changes in soil pH can be slow similar to other soil properties. Reducing soil acidity with cover crops may take more time than with lime application, but it can reduce the amount of lime needed in the long term.

The variable impacts of cover crops on soil pH can depend on soil texture, initial soil pH level, and climate. First, cover crops generally increase soil pH in degraded soils such as those with low pH (Obi, 1999), low fertility (Santos et al., 2022), highly eroded conditions (Rankoth et al., 2019), and high sand content (Khan et al., 2021). Indeed, Table 11.2 shows that when soil pH is nearly neutral (around 7), cover crops have limited or no effects on soil pH. Also, cover crops appear to improve soil pH more in tropical soils than in temperate soils (Santos et al., 2022; Obi, 1999). Research data on soil pH and other soil properties support the argument that degraded soils can preferentially benefit from cover crop introduction. Changes in soil pH may also depend on differences in biomass production although studies correlating changes in soil pH with cover crop biomass production are unavailable. Note that time after cover crop introduction does not appear to have large effects (Table 11.2). Overall, cover crops offer potential to manage soil pH in acidic and degraded soils, but, in temperate or highly productive soils, changes in soil pH with cover crops may be relatively small and not measurable in the short term.

Table 11.2 Cover Crop Impacts on Soil pH Across Different Environments in the Upper 30 cm Depth

Location	Soil	Initial soil pH	Cover crops	Duration (yr)	Soil pH
[a]New Mexico, U.S.	Clay loam	7.30	Pea, oat, canola, and their mixes	5	No effect
[b]New York, U.S.	Silt loam	7.86	Ryegrass, red clover, crimson clover, and hairy vetch	4	No effect
[c]Nebraska, U.S.	Silty clay loam	7.10	Cereal rye, grain sorghum, Austrian winter pea, and soybean	12	No effect

(Continued)

Table 11.2 (Continued)

Location	Soil	Initial soil pH	Cover crops	Duration (yr)	Soil pH
[d]Foulum, Denmark	Sandy loam	5.97	Radish	5	No effect
[e]Nsukka, Nigeria	Sandy clay loam (acidic)	4.40	Carpet grass, creeping grass, guinea grass, elephant grass, style, and Kudzu	5	Creeping grass and guinea grass increased to 4.6; Others had no effect
[e]Piaui State, Brazil	Sandy clay loam (degraded)	5.10	Multiple single species and mixes	3	Increased to 5.90
[f]Missouri, U.S.	Loam (eroded)	5.52	Mix of winter cover crops (rye, barley, triticale, and oat)	6	Increased to 5.78
[g]Prince Edward Island, Canada	Sandy loam	6.30	Timothy, buckwheat, brown mustard, and mix (winter rye, hairy vetch, and crimson clover)	1	Brown mustard increased to 6.82; Others had no effect

[a] Thapa et al. (2022).
[b] Nunes et al. (2018).
[c] Blanco-Canqui and Jasa (2019).
[d] Abdollahi and Munkholm (2014).
[e] Obi (1999).
[f] Santos et al. (2022).
[g] Rankoth et al. (2019), Khan et al. (2021).

11.6 Cation Exchange Capacity

It is well known that cation exchange capacity is a critical soil fertility property that determines the capacity of the soil to adsorb, retain, exchange, and buffer positively-charged ions including K^+, Ca^{++}, Mg^{++}, Na^+, NH_4^+, and others. Similar to soil pH, cation exchange capacity directly affects the nutrient availability in the soil solution, thereby influencing the ability of the soil to readily supply nutrients to plants. Despite their importance, research data on the impact of cover crops on this soil property are, however, sparse. It is commonly believed that cover crops increase the cation exchange capacity of the soil (Dabney et al., 2001), but what does research show?

Available studies summarized in Table 11.2 indicate that cover crop impacts on cation exchange capacity can be mixed. Cover crops may increase, reduce, or have no effect on cation exchange capacity. The capacity of the soil to exchange cations is often positively correlated with soil organic matter concentration. However, in some cases, even a significant increase in organic matter concentration with cover crops does not correspond to increased cation exchange capacity. Differences in soil texture, initial soil organic matter level, and climate can influence cover crop effects on cation exchange capacity (Table 11.2). First, cover crops do not appear to rapidly change cation exchange capacity when the initial organic matter level of the soil is high (Blanco-Canqui and Jasa, 2019). Second, cover crops could rapidly alter cation exchange capacity in sandy or coarse-textured soils (dos Santos Cordeiro et al., 2021; Obi, 1999). Third, while data are relatively few, cover crops can alter cation exchange capacity more readily in tropical than in temperate soils (Blanco-Canqui and Jasa, 2019; Santos et al., 2022).

The decrease in cation exchange capacity after cover crop introduction, in a few cases, can be due to the positive priming effect of cover crops on organic matter (Santos et al., 2022). Because organic matter is related to cation exchange capacity, cation exchange capacity can decrease when addition of fresh cover crop residues boosts mineralization of the orginal organic matter and reduce soil organic matter concentration (positive priming). Indeed, as discussed earlier, while cover crops increase soil organic matter concentration in most cases, it can reduce it in a few cases due to the positive priming (Duval et al., 2016; Santos et al., 2022). Research suggests that cover crops may not always improve the ability of the soil to retain and exchange cations. Cover crops can alter cation exchange capacity more rapidly in low organic matter, sandy, and tropical soils than in high organic matter, high clay content, and temperate soils.

11.7 Carbon to Nitrogen Ratio

Cover crops release nutrients via residue decomposition after termination. Some of the factors that affect the rate of cover crop residue mineralization include quality (e.g., C to N ratio, C to P ratio, forms of C) of cover crop residues, soil texture, biological activity, and climatic conditions. In general, the release of nutrients from cover crop residues increases with an increase in residue quality (e.g., high N concentration), biological activity (e.g., earthworm activity), precipitation amount, temperature, and sand content of the soil. In a tropical region, hairy vetch released as much as 90% of total N within four weeks after termination (Acosta et al., 2011).

Legume cover crops, due to their lower C to N ratio, decompose faster than non-legume cover crops. The ratio of C to N can differ even among legume cover crop species. Claassen (2009) reported sunn hemp cover crop residues had $23.7\,g\,kg^{-1}$ of N, but late-maturing soybean residues had $31.1\,g\,kg^{-1}$ of N, which suggests that soybean cover crop residues could decompose faster and provide N more rapidly than sunn hemp residues. Synchronizing the release of N from legume cover crops with the peak uptake by main crops is key to reduce N fertilizer requirements. Hairy vetch, one of the most common legume cover crops, can replace significant amounts of mineral N fertilizer. For example, Ketterings et al. (2015) found that the "N fertilizer replacement value" for hairy vetch ranged from 17 to $149\,kg\,N\,ha^{-1}$.

Cover crop residues with a C to N ratio below 20 and C to P ratio below 300 mineralize more rapidly than those with a C to N ratio above 20 and C to P ratio above 300. For example, the C to N ratio can be approximately $35:1$ for winter rye and $10:1$ for hairy vetch, but the ratio within each species depends on the growth stage (Sievers & Cook, 2018). Soil N can be immobilized if the C to N ratio is high such as under grass cover crops. Early-terminated cover crops can have a lower C to N ratio than when the same cover crop is terminated late as younger plants have higher N concentration and are more easily decomposable. In general, legume cover crops rapidly release N in the first three or four weeks after termination, but grass cover crops release N more slowly. Thus, grass cover crops can behave like slow-release N fertilizer, supplying N gradually as residues decompose. Determining the amount of nutrient released from a cover crop is important to reduce the amount of inorganic fertilizer needed by the next crop. Also, matching the nutrient release with the period when crops need nutrients the most can be key to reduce N losses. It is estimated that crops will utilize only approximately 30% of N released from legume cover crops.

A related question is: What is the amount of plant available N for the subsequent crops after cover crop termination? The exact amount of available N can be difficult to quantify due to many interacting factors including the C to N ratio of cover crop residues, soil type (e.g., clayey, sandy), and climate (e.g, temperate, tropical). In general, it is estimated that approximately 5–30% of the total N can become available for the next crop, depending on the C to N ratio. Indeed, the C to N ratio of cover crop residues is one of the critical factors that determine cover crop residue decomposition and N availability. The higher the C to N ratio, the lower the amount of available N for the subsequent crop. For instance, only approximately 5% of total N may be plant available N if the C to N ratio of cover crop residues is approximately $30:1$, while approximately 30% of total N may be plant available N if the C to N ratio of cover crop residues is approximately $10:1$. This example highlights the critical importance of the C to N ratio to N availability.

11.8 Summary

Cover crops increase soil organic matter concentration in most cases and this increase appears to be high in soils with low initial organic matter. Also, cover crops such as summer legume cover crops can produce high amounts of N-enriched biomass in warm climates, potentially reducing the need for N fertilizer input for the following crop. Furthermore, brassica cover crops can release larger amounts of P than other cover crops.

Cover crops can improve P efficiency by reducing losses of P associated with sediment. However, while cover crops retain most P in the soil, they could increase losses of dissolved P in runoff. This indicates that cover crops can be a source and sink of P, which warrant a site-specific consideration of tradeoffs when using cover crops for soil P management. Grass cover crops can primarily contribute to soil fertility by scavenging nutrients and reducing their losses, whereas legume cover crops can directly increase soil fertility by processing atmospheric N.

Introduction of cover crops may not improve pH and cation exchange capacity in highly fertile soils but can improve these soil fertility indicators in degraded soils including acidic, sandy, eroded, and low organic matter soils. Non-legume cover crops can improve soil fertility by reducing erosion and N emissions. Development of site-specific cover crop management strategies is key to optimize cover crop benefits for nutrient management. Overall, cover crops improve soil organic matter concentration, retain and release essential nutrients (N, P, K, Ca, Mg, and others), and, in some cases, improve soil pH and cation exchange capacity.

References

Abdollahi, L., & Munkholm, L. J. (2014). Tillage system and cover crop effects on soil quality: I. Chemical, mechanical, and biological properties. *Soil Science Society of America Journal, 78*, 262–270.

Acosta, J. A. A., Amado, T. J. C., Neergaard, A., Vinther, M., Silva, L. S., & Nicoloso, R. S. (2011). Effect of 15N-labeled hairy vetch and nitrogen fertilization on maize nutrition and yield under no-tillage. *Revista Brasileira de Ciência do Solo, 35*, 1337–1345.

Basche, A. D., Miguez, F. E., Kaspar, T. C., & Castellano, M. J. (2014). Do cover crops increase or decrease nitrous oxide emissions? A meta-analysis. *Journal of Soil and Water Conservation, 69*, 471–482.

Bennett, H. H. (1939). *Soil conservation*. McGraw-Hill Book Co., Inc. 993 pages.

Blanco-Canqui, H. (2018). Cover crops and water quality. *Agronomy Journal, 110,* 1633–1647.

Blanco-Canqui, H., Claassen, M. M., & Presley, D. R. (2012). Summer cover crops fix nitrogen, increase crop yield and improve soil-crop relationships. *Agronomy Journal, 104,* 137–147.

Blanco-Canqui, H., & Jasa, P. (2019). Do grass and legume cover crops improve soil properties in the long term? *Soil Science Society of America Journal, 83,* 1181–1187.

Blanco-Canqui, H., Ruis, S., Holman, H., Creech, C., & Obour, A. (2022). Can cover crops improve soil ecosystem services in water-limited environments? A review. *Soil Science Society of America Journal, 86,* 1–18.

Blanco-Canqui, H., Ruis, S., Proctor, C., Creech, C., Drewnoski, M., & Redfearn, D. (2020). Harvesting cover crops for biofuel and livestock production: Another ecosystem service? *Agronomy Journal, 112,* 2373–2400.

Carver, R. E., Nelson, N. O., Roozeboom, K. L., Kluitenberg, G. J., Tomlinson, P. J., Kang, Q., & Abel, D. S. (2022). Cover crop and phosphorus fertilizer management impacts on surface water quality from a no-till corn-soybean rotation. *Journal of Environmental Management, 301,* 113818.

Claassen, M. M. (2009). Effects of late-maturing soybean and sunn hemp summer cover crops and nitrogen rate in a no-till wheat/grain sorghum rotation. *Report of Progress, 1017,* 44–49. Available at http://www.ksre.ksu.edu/library/crpsl2/srp1017.pdf

Dabney, S. M., Delgado, J. A., & Reeves, D. W. (2001). Using winter cover crops to improve soil and water quality. *Communications in Soil Science and Plant Analysis, 32,* 1221–1250.

dos Santos Cordeiro, C. F., Echer, F. R., & Araujo, F. F. (2021). Cover crops impact crops yields by improving microbiological activity and fertility in sandy soil. *Journal of Soil Science and Plant Nutrition, 21,* 1968–1977.

Duval, M. E., Galantini, J. A., Capurrod, J. E., & Martinez, J. M. (2016). Winter cover crops in soybean monoculture: Effects on soil organic carbon and its fractions. *Soil and Tillage Research, 161,* 95–105.

Groff, S. (2015). The past, present, and future of the cover crop industry. *Journal of Soil and Water Conservation, 70,* 130.

Hallama, M., Pekrun, C., Lambers, H., & Kandeler, E. (2019). Hidden miners - the roles of cover crops and soil microorganisms in phosphorus cycling through agroecosystems. *Plant Soil, 434,* 7–45.

Kasper, S., Christoffersen, B., Soti, P., & Racelis, A. (2019). Abiotic and biotic limitations to nodulation by leguminous cover crops in South Texas. *Agriculture, 9,* 209.

Ketterings, Q. M., Swink, S. N., Duiker, S. W., Czymmek, K. J., Beegle, D. B., & Cox, W. J. (2015). Integrating cover crops for nitrogen management in corn systems on northeastern U.S. dairies. *Agronomy Journal, 107,* 1365–1376.

Khan, R., Farooque, A. A., Brown, H. C. P., Zaman, Q. U., Acharya, B., Abbas, F., & McKenzie-Gopsill, A. (2021). The role of cover crop types and residue incorporation in improving soil chemical properties. *Agronomy, 11*, 2091.

Kim, N., Zabaloy, M. C., Guan, K., & Villamil, M. B. (2020). Do cover crops benefit soil microbiome? A meta-analysis of current research. *Soil Biology and Biochemistry, 142*, 107701.

Liu, J., Macrae, M. L., Elliott, J. A., Baulch, H. M., Wilson, H. F., & Kleinman, P. J. A. (2019). Impacts of cover crops and crop residues on phosphorus losses in cold climates: A review. *Journal of Environmental Quality, 48*, 850–868.

Lozier, T. M., Macrae, M., Brunke, R., & Van Eerd, L. L. (2017). Release of phosphorus from crop residue and cover crops over the non-growing season in a cool temperate region. *Agricultural Water Management, 189*, 39–51. https://doi.org/10.1016/j.agwat.2017.04.015

Mansoer, Z., Reeves, D. W., & Wood, C. W. (1997). Suitability of sunn hemp as an alternative late-summer legume cover crop. *Soil Science Society of America Journal, 61*, 246–253.

Nunes, M. R., van Es, H. M., Shindelbeck, R., Ristow, A. J., & Ryan, M. (2018). No-till and cropping system diversification improve soil health and crop yield. *Geoderma, 328*, 30–43.

Obi, M. E. (1999). The physical and chemical responses of a degraded sandy clay loam soil to cover crops in southern Nigeria. *Plant and Soil, 211*, 165–172.

Poeplau, C., & Don, A. (2015). Carbon sequestration in agricultural soils via cultivation of cover crops-a meta-analysis. *Agriculture, Ecosystems and Environment, 220*, 33–41.

Rankoth, L. M., Udawatta, R. P., & Gantzer, C. J. (2019). Cover crops on temporal and spatial variations in soil microbial communities by phospholipid fatty acid profiling. *Agronomy Journal, 111*, 1693–1703.

Santos, G. G., Rosetto, S. C., Barbosa, R. S., Melo, N. B., Soares de Moura, M. C., Santos, D. P., Flores, R. A., & Collier, L. S. (2022). Are chemical properties of the soil influenced by cover crops in the Cerrado/Caatinga ecotone? *Communications in Soil Science and Plant Analysis, 53*, 89–103.

Sievers, T., & Cook, R. L. (2018). Aboveground and root decomposition of cereal rye and hairy vetch cover crops. *Soil Science Society of America Journal, 82*, 147–155.

Thapa, R., Mirsky, S. B., & Tully, K. L. (2018). Cover crops reduce nitrate leaching in agroecosystems: A global meta-analysis. *Journal of Environmental Quality, 47*, 1400–1411.

Thapa, V. R., Ghimire, R., VanLeeuwen, D., Acosta-Martinez, V., & Shukla, M. K. (2022). Response of soil organic matter to cover cropping in water-limited environments. *Geoderma, 406*, 115497.

Tonitto, C., David, M. B., & Drinkwater, L. E. (2006). Replacing bare fallows with cover crops in fertilizer-intensive cropping systems: A meta-analysis of crop yield and N dynamics. *Agriculture, Ecosystems and Environment, 112*, 58–72.

Varela, M. F., Barraco, M., Gili, A., Taboada, M. A., & Rubio, G. (2017). Biomass decomposition and phosphorus release from residues of cover crops under no-tillage. *Agronomy Journal, 109*, 317–326.

Villamil, M. B., Bollero, G. A., Darmody, R. G., Simmons, F. W., & Bullock, D. G. (2006). No-till corn/soybean systems including winter cover crops: Effects on soil properties. *Soil Science Society of America Journal, 70*, 1936–1944.

Weiler, D. A., Giacomini, S. J., Aita, C., Schmatz, R., Pilecco, G. E., Chaves, B., & Bastos, L. M. (2019). Summer cover crops shoot decomposition and nitrogen release in a no-tilled sandy soil. *Revista Brasileira de Ciência do Solo, 43*, e0190027.

12

Crop Yields

12.1 Multi-functionality of Cover Crops

Cover crops generally reduce soil erosion, suppress weeds, reduce nutrient leaching, maintain soil fertility, and improve soil physical properties, soil microbiome, and C and nutrient cycling, and provide other soil ecosystem services as discussed in the previous chapters. The question is: Do the above benefits from cover crops translate into increased crop yields compared with conventional systems without cover crops? The answer to this question may not be simple. It is important to remember that cover crops were not initially conceived to boost crop yields but primarily to cover the soil, reduce erosion, and maintain soil fertility.

Moreover, cover crops were originally introduced into low-input agricultural systems with limited or no synthetic fertilizer application (Bennett, 1939). The introduction of synthetic fertilizers in the early to middle twentieth century resulted in large increases in crop yields relative to low-input systems. Now, interest in cover crops is reemerging to address the growing agronomic and environmental challenges, but it is important to consider that cover crops may be more effective for addressing soil erosion, leaching, non-point source water pollution, and other environmental concerns than for increasing crop yields to levels similar to those under high-input conventional systems.

Implementing conservation practices such as cover crops to address soil and environmental challenges is commendable, but some can be hesitant to adopt cover crops if impact on crop yields is negative. Concerns exist that cover crops can reduce subsequent crop yields by depleting water for the following crop in water-limited regions and immobilizing (e.g., non-legume cover crops) nutrients. Additionally, in no-till soils, cover crop residue mulch after late termination could delay soil warming, reduce seed–soil contact and seed germination, and thus

---

Cover Crops and Soil Ecosystem Services, First Edition. Humberto Blanco.
© 2023 American Society of Agronomy, Inc. / Crop Science Society of America, Inc.
/ Soil Science Society of America, Inc. Published 2023 by John Wiley & Sons, Inc.

adversely affect crop production. This chapter discusses how cover crops affect crop production based on the available research information.

12.2 Crop Yields

According to two global reviews, cover crops either have no effect or reduce subsequent crop yields in most cases (Table 12.1). The first global review indicated that cover crops can reduce crop yields by approximately 4% relative to no cover crops (Abdalla et al., 2019; Table 12.1). The second review concluded that non-legume cover crops did not affect crop yields when crops received inorganic fertilizer at recommended rates compared with no cover crops (Tonitto et al., 2006). The same review concluded that crops fertilized with legume cover crops had 10% lower yields compared with crops fertilized with inorganic N fertilizer.

While the above reviews suggest cover crops may not increase crop yields, the reported percent crop yield reduction is an average number across all reviewed studies. In some locations or regions, cover crops such as legumes or mixes dominated by legumes can increase crop yields relative to no cover crops (Marcillo & Miguez, 2017). Factors including climate, cover crop species (legumes versus non-legumes), cover crop biomass production, tillage and cropping systems, inorganic fertilization, cover crop planting and termination dates, climate (e.g., cool and warm temperate regions), and their interactions affect the magnitude of cover

Table 12.1 Synthesis of Review Papers on Cover Crops and Crop Yields on Regional and Global Scales

Regions	Number of studies	Cover crop effect on yield
[a]Semiarid	20	No effect in 56% of comparisons; Reduced in 38% of comparisons; Increased in 6% of comparisons
[b]Temperate (U.S. and Canada)	65	Grass winter cover crops had no effect on corn yield; Legume cover crops increased corn yields by 30–33% when N fertilization was low
[c]Across All Regions	35	Non-legume cover crops did not reduce yields of crops receiving fertilizer at recommended rates; Crop fertilized with legume cover crop had 10% lower yields compared with that fertilized with inorganic N
[d]Across All Regions	40	Cover crops reduced grain yield by ~4%

[a] Blanco-Canqui et al. (2022).
[b] Marcillo and Miguez (2017).
[c] Tonitto et al. (2006).
[d] Abdalla et al. (2019).

crop impacts (Deines et al., 2023). How cover crops affect subsequent crop yields can be complex. Thus, cover crop impacts on crop yields deserve an examination on a site-specific basis, considering all the crop production factors.

12.3 Climate

12.3.1 Cool and Warm Climates

Cover crop impacts on crop yields can vary between cool and warm climates. A regional review of cover crops focused on corn yields across studies in the U.S. and Canada concluded the impact of winter cover crops on corn yields varied with climatic region. Winter cover crops had no effect on corn yield relative to no cover crops in the Great Plains, Canada, and U.S. North Central region but increased corn yields by 12–14% in U.S. Southeast, Northeast, and Northwest when inorganic N fertilizer rate was zero or lower than the recommended N rates (Marcillo & Miguez, 2017). In cool temperate regions (e.g., Canada, Northern U.S.), winter cover crops such as legumes can have limited positive effects on corn yields due to limited biomass production from short growing window for cover crops. A field-scale analysis across the U.S. Corn Belt estimated that cover crops reduced corn yield by 5.5% on croplands with cover crops > 3 yr (Deines et al., 2023). Further, corn yield loss was larger on *"fields with better soil ratings, cooler mid-season temperatures, and lower spring rainfall"*. The same study found that cover crops reduced soybean yield by 3.5%, *"with larger impacts on fields with warmer June temperatures, lower spring and late-season rainfall, and, to a lesser extent, better soils"*.

When biomass production is limited, benefits of cover crops for reducing water erosion, nitrate leaching, N_2O emissions, non-point source pollution of water, and addressing other environmental concerns can also be small. However, in warm temperate or humid subtropical regions with mild winters, winter cover crops such as legumes can grow longer and produce more N-enriched biomass for boosting subsequent corn yields than in regions with long and cold winters. Winter cover crops could more rapidly decompose and release nutrient in warmer regions than in cool temperate regions. This corroborates cover crop performance is region- or climate-dependent.

12.3.2 Water-Limited Regions

It is often presumed cover crops can reduce crop yields in arid and semiarid regions by using water needed for the next crop. A review of cover crops in water-limited regions (<500 mm precipitation) reported that cover crops do not affect crop yields in 56% of comparisons, reduce in 38%, and increase in 6% (Blanco-Canqui et al., 2022). This indicates cover crops do not always reduce crop yields in water-limited regions. Also, the same review found cover crops can reduce soil water in only approximately 50% of cases, which mirrors the mixed effects of cover crops on crop yields. A recent study in the southeastern U.S. Great Plains reported pea, oat, and canola cover crops and their mixes had mixed effects on crop yields during a five-year study, indicating cover crops do not consistently reduce subsequent crop yields in water-limited environments (Acharya et al., 2022). Cover crops can reduce subsequent crop yields when precipitation amount is below normal or no precipitation falls between cover crop termination and main crop planting (Holman et al., 2018; Nielsen et al., 2015). In water-limited regions, cover crops are sometimes irrigated to ensure establishment and sufficient biomass production, but irrigation could mask the benefits of cover crop residue to conserve water and contribute to crop production.

12.4 Factors Affecting Crop Production

12.4.1 Cover Crop Species

Cover crop effects on crop production can particularly depend on which cover crop species are used. Non-legume cover crops such as grasses could reduce or have no effect on subsequent crop yields by immobilizing nutrients or competing for nutrients with the main crop. However, legume cover crops can increase subsequent crop yields by fixing atmospheric N and improving soil fertility. Rapid decomposition of legume residues relative to grass cover crops can also reduce residue mulch effects on soil temperature or interference with crop planting. Experimental data suggest that legume cover crops offer potential for increasing corn yields while providing other soil ecosystem services. Performance of legume cover crops is closely linked with climate. Legumes can increase crop yields particularly in relatively warm climates due to increased biomass production and residue mineralization for N provision. However, benefits from legume cover crops for crop yields may not be observed in all cases, especially when biomass production is low (Marcillo & Miguez, 2017).

It is important to stress that while non-legume cover crops do not increase crop yields, they can provide other essential ecosystem services not directly related to crop yields including water and wind erosion control, weed suppression, improved water quality, reduced N_2O emissions, and soil C sequestration (Blanco-Canqui et al., 2015; Schipanski et al., 2014). One may expect that non-legume cover crops can rapidly increase crop yields by suppressing weeds, reducing soil compaction, and improving other factors of crop production, but research data indicate these positive services from non-legume cover crops do not necessarily manifest in increased yields of subsequent crops compared with no cover crops in all cases.

12.4.2 Nitrogen Fertilization

Cover crops interact with inorganic fertilization to impact subsequent crop yields. Legume cover crops often increase crop yields when inorganic N fertilizer application rate is zero or low. In a review, Tonitto et al. (2006) found that crops fertilized with legume cover crops had 5% higher yield compared with those fertilized with low rates of inorganic fertilizer. Also, a long-term study of legume crops (sunn hemp and late-maturing soybean) under no-till winter wheat-sorghum in Kansas found that cover crop benefits on crop yields decreased as N fertilizer application rates increased from 0 to 100 kg N ha^{-1} (Blanco-Canqui et al., 2012). For example, cover crops increased sorghum yields the most (36.7%) at 0 kg N ha^{-1} of fertilizer application. Similarly, a review of all studies across U.S. and Canada reported that non-legume cover crops increased corn yields when corn was fertilized with low amounts of N (Marcillo & Miguez, 2017). Compared with no cover crop, legume cover crops increased crop yields by approximately 30% when N rate was <100 kg N ha^{-1} and by 9% when the N rate was between 100 and 199 kg N ha^{-1}, but had no effect on corn yields when the N rate was 200 kg N ha^{-1}.

Research indicates that application of inorganic fertilizer at high rates can mask the benefits of legume cover crops. The improved soil properties and processes with cover crops may not be sufficient to overcome increases in crop yields under high input systems (e.g., high inorganic fertilization). Also, impacts of cover crops on subsequent crop yields can depend on the main crop type. Compared with systems with inorganic fertilizers, Tonitto et al. (2006) discussed legume cover crops reduced corn yields by 12% but not sorghum and vegetable yields. Crops fertilized with high rates of inorganic N can outperform crops fertilized only with legume cover crops. Thus, experimental data suggest legume cover crops can be more beneficial in systems with limited fertilizer input or in organic systems than in systems receiving high amounts of inorganic fertilizers. Yet, it is still beneficial to introduce cover crops in high-input systems. Under high-input systems, legume cover crops may not completely replace inorganic fertilizer but can reduce the amount of fertilizer required for the crops.

12.4.3 Biomass Production

The implications of an increase in cover crop biomass production on crop yields can depend on cover crop species. An increase in legume cover crop biomass production can increase crop yields by releasing significant amounts of N under favorable climatic and management conditions. For instance, summer legume cover crops can produce large amounts of biomass and thus N for the subsequent crops. A study in Alabama found that sunn hemp produced 7.6 Mg ha^{-1} of biomass with 144 kg N ha^{-1} and increased corn yield by 1.2 Mg ha^{-1} compared with no cover crops in two of three years (Balkcom & Reeves, 2005). Also, in Kansas, sunn hemp and late-maturing soybean produced more than 6 Mg ha^{-1} of biomass across multiple years and increased crop yields particularly under low or no inorganic N fertilization (Blanco-Canqui et al., 2012). These results indicate high-biomass producing legume cover crops can add N and reduce N fertilizer requirements.

However, an increase in grass cover crop biomass production can lead to immobilization of nutrients and reduce subsequent crop yields. Also, after termination, grass cover crop residue mulch in no-till systems can reduce soil temperature in spring, interfere with planting, reduce soil–seed contact, and reduce germination, potentially reducing crop yields. For example, late-terminated grass cover crops can produce significant amounts of biomass and result in reduced crop yields in some years (Ruis et al., 2017). Also, it is important to note that high biomass production can use more water than low biomass production regardless of cover crop species, which may thus reduce yields of the subsequent crop in water-limited regions.

12.4.4 Planting Time and Method

Do cover crop planting time and planting method affect subsequent crop yields? This question has been subject of recent research as many seek strategies or opportunities to succeed with cover crop management. Particularly in cool temperate regions, winter cover crops such as winter rye drilled in late fall after corn or soybean harvest and terminated in early spring several weeks before crop planting often produce low amounts biomass with limited benefits (Ruis et al., 2019). Interseeding cover crops in crops can be an alternative to post-harvest planting for increasing growing window of cover crops although interseeded cover crops do not often produce more biomass than post-harvest planted cover crops (Ruis et al., 2020; Stanton & Haramoto, 2021). In some cases, interseeded cover crops do not establish well due to shading from main crops, limited rainfall after interseeding, poor soil–seed contact, weed competition, and slow emergence.

Research data show interseeded cover crops do not impact subsequent crop yields, which is similar to the impacts of post-harvest planted cover crops (Table 12.2). In a few cases, interseeded cover crops can reduce crop yields, depending on cover crop termination date or year (Curran et al., 2018; Koehler-Cole et al., 2020; Noland et al., 2018). Late termination of the interseeded cover crops can produce more biomass than early termination, although it can reduce crop yields in some years. Delaying termination can be beneficial for

Table 12.2 Some Case Studies of Interseeding Cover Crops and Crop Yield Response in Temperate Regions

Location	Crop	Cover crop	Planting time before crop harvest	Years	Interseeding effect on yield
[a]Three sites in Nebraska, U.S.	Soybean	Cereal rye and mix (rye, legume, and brassica)	~1 mo	4	Limited or no effect
[b]Two sites in Minnesota, U.S.	Corn–soybean	Winter rye, red clover, hairy vetch, field pennycress, and mix (oat, pea, and tillage radish)	~4 mo	3	Limited or no effect
[c]New York, Pennsylvania, and Maryland, U.S.	Corn	Annual ryegrass, red clover, crimson clover, and hairy vetch	3–4 mo	1	No effect
[d]Two sites in Michigan, U.S.	Corn	Ryegrass, crimson clover, oilseed radish, and mix	~4 (late) and 5 (early) months	2	No effect
[e]Minnesota, U.S.	Sugar beet	Austrian pea, winter camelina, brown mustard, and winter rye	~92 (early) and 79 (late) days	3	No effect

[a] Koehler-Cole et al. (2020)
[b] Noland et al. (2018)
[c] Curran et al. (2018).
[d] Brooker et al. (2020).
[e] Sigdel et al. (2021).

reducing nitrate leaching, suppressing weeds, and increasing potential for soil C accumulation.

Research also indicates that methods of seeding cover crops do not generally affect crop yields although long-term and comprehensive studies comparing different methods (e.g., drilling, aerial broadcasting, surface broadcasting) within the same experiment are still few. Drilling cover crops often produces more cover crop biomass than broadcasting due to better soil–seed contact, as discussed earlier, but such planting methods may not affect cover crop effects on subsequent crop yields. A study in the upper U.S. Midwest found that corn and soybean yields among three cover crop planting methods (broadcast, broadcast with light incorporation, and a high-clearance drill) did not generally differ (Noland et al., 2018). However, the same study found that drilled cover crops generally produced more biomass than the broadcast method, which can be important for enhancing soil ecosystem services other than crop yields. Interseeding cover crops has limited or no effect on subsequent crop yields in most cases.

12.4.5 Termination Timing

Particularly in water-limited regions, early termination of cover crops can be recommended to reduce water use by cover crop for the subsequent crop. Early termination can allow water recharge between termination and main crop planting. In water-limited regions, cover crops can replace fallow periods in crop-fallow systems. However, the reduction in soil water and crop yields in approximately 50% of cases in these systems (Blanco-Canqui et al., 2022) indicates that cover crops should be grown only during a portion of fallow to limit water depletion. In the eastern U.S. Great Plains, Ruis et al. (2017) found cover crops terminated approximately 30 days before corn planting did not affect corn yield in a three-year study, but cover crops terminated at corn planting (late termination) reduced corn yield in one of three years by 8% (from 16.8 to 15.4 Mg ha^{-1}).

Late termination of cover crops can be detrimental in years when precipitation is low during crop establishment. It can also reduce yields due to high cover crop residue input, which can reduce soil temperature and thus crop germination. Studies evaluating how cover crop termination timing affects crop yields under different climates and management scenarios are few. Late-terminated cover crops may reduce crop yields in low precipitation years but can provide other essential ecosystem services including greater weed suppression, erosion control, C accumulation, and cover crop biomass production for expanded uses, among others. Flexible cover crop termination dates can be a strategy to adjust to varying precipitation input from year to year and thus reduce potential adverse effects on crop yields.

12.4.6 Cover Crop Mixes

Much enthusiasm exists about introducing cover crop mixes into cropping systems to promote plant diversity and potentially enhancing soil ecosystem services. Because plant species could have different resource requirements and response to environmental conditions (Chapagain et al., 2020; Tamburini et al., 2020), some consider that mixing different cover crop species could better utilize resources and adapt to changing weather or environmental conditions compared with single species. Thus, the question is: Do multi-species cover crop mixes increase crop yields relative to single species cover crops?

Reviews of published data indicate that cover crop mixes do not increase crop yields over single species in most cases (Florence & McGuire, 2020). A global review by Florence and McGuire (2020) concluded that cover crop mixes had no effect on crop yields in most comparisons compared with single species cover crops. The limited or lack of superiority of mixes over single species for increasing crop yields is expected because mixes, as discussed in earlier chapters, do not outperform single species for producing biomass, storing soil water, providing nutrients, suppressing weeds, pests, and diseases, and enhancing other soil ecosystem services that affect crop production.

It is important to note that in a few cases where mixes increase crop yields more than monocultures, such increase may be simply due to differences in seeding rate. The review by Florence and McGuire (2020) found that cover crop mix increased crop yields in one of 14 comparisons, but such increase occurred when the mix was seeded at a higher seeding rate than monocultures. Differences in seeding rate for the same cover crop species when grown in a mix and when grown alone can mask the true impacts of mixes on crop yields. Overall, cover crop mixes do not generally increase crop yields compared with monocultures.

12.4.7 Years after Cover Crop Adoption

Cover crop biomass production following adoption could determine how fast cover crops can affect subsequent crop yields. Cover crops such as legumes can increase crop yields even in the first year if they accumulate significant amounts of N-enriched biomass. Blanco-Canqui et al. (2012) found sunn hemp cover crop increased sorghum yield by $0.57\,Mg\,ha^{-1}$ in the first year of introduction when sunn hemp produced $8.81\,Mg\,ha^{-1}$ of biomass. In general, the extent to which legume cover crops increase crop yields can increase with years after cover crop introduction as soil N accumulates and other soil properties (e.g., organic matter) improve with time.

Differences in crop yields between cover crop and no cover crop can increase gradually with time, especially when no N fertilizer was applied. However, cover crops that produce low amounts of biomass ($<1\,Mg\,ha^{-1}$) may have limited effects on crop yields even after several years of cover crop use. For example, in the eastern U.S. Great Plains, winter rye cover crop terminated approximately two or three weeks before corn planting produced $0.03\,Mg\,ha^{-1}$ of biomass in the first year, $0.08\,Mg\,ha^{-1}$ in the second year, $1.41\,Mg\,ha^{-1}$ in the third year and did not reduce subsequent yields in a rainfed no-till continuous corn (Ruis et al., 2017). Years after cover crop introduction may have smaller impacts on crop yields than differences in cover crop biomass production.

12.4.8 Tillage Systems

Tillage systems can differently affect cover crop effects on crop yields by differently impacting water conservation, cover crop residue decomposition, and other processes of crop production. Marcillo and Miguez (2017), in a review, found legume cover crops increased corn yields by 15% in tilled systems and by 30% in no-till systems. Similarly, in a review, Abdalla et al. (2019) concluded cover crops reduced crop yields in tilled systems but not in no-till, reduced till, and minimum till systems. Experimental data thus indicate cover crops can perform better when combined with no-till than when combined with tilled systems.

No-till management could enhance cover crop performance by a number of mechanisms (Marcillo & Miguez, 2017). One, cover crop residues left on the surface in no-till soils after termination can reduce evaporation and thus maintain soil water relative to tilled systems (Blanco-Canqui et al., 2012). This is particularly important in water-limited rainfed systems. Two, cover crop residues, even legumes, can decompose slower in no-till than in tilled systems. As result, no-till management may better synchronize nutrient release with crop nutrient uptake (Abdalla et al., 2019). Tilling cover crops as green manure can accelerate residue mineralization and facilitate rapid nutrient release but can also accelerate nutrient losses (e.g., gas emissions, leaching) before crop establishment. Cover crops such as legumes combined with no-till may be a better option for maintaining or increasing crop yields than combined with tilled systems.

12.4.9 Soil Texture

Coarse-textured and low-organic matter soils can benefit more from cover crops than highly fertile or clayey soils. Indeed, research suggests that cover crops such as legumes can boost crop yields more in erosion-prone and coarse-textured soils than in cool and clayey soils although N losses via leaching can be a concern in coarse-textured soils (Marcillo & Miguez, 2017). In warm climates and coarse-textured soils, legume cover crops can decompose rapidly and provide much needed N to subsequent crops. Available data suggest, however, that the influence of soil texture on the magnitude of cover crop impacts on crop yields can be smaller than the influence of cover crop species, biomass production, and climate. For example, Tonitto et al. (2006), in a global review, found that differences in soil texture had limited influence on cover crop effects on yields. Also, a review by Blanco-Canqui et al. (2022) focusing on water-limited regions reported that differences in soil textural classes had minimal or no effects on cover crop impacts on crop yields.

12.5 Summary

Cover crops increase, reduce, or have no effect on crop yields. Cover crops improve many soil ecosystem services but such improvement does not appear to translate into increased crop yields in all management or climatic scenarios. Similar to other ecosystem services, cover crop impacts on crop yields can be highly site- or context-specific. Differences in cover crop species, inorganic N fertilizer rate, cover crop growing season, amount of biomass produced, and climate can explain the inconsistent crop yield response to cover crop.

Non-legume cover crops can maintain or reduce crop yields, which indicates that they can be mainly used for reducing erosion and leaching, suppressing weeds, accumulating C, and other soil services. Legume cover crops, in turn, can increase crop yields under zero or low inorganic N fertilizer rate. The positive impact of legumes can be greater in warmer temperate than in cool temperate regions due to longer growing cover crop season in warm climates. Cover crops do not reduce subsequent crop yields in approximately 50% of cases in water-limited regions. Also, planting early or interseeding cover crops do not outperform post-harvest planted cover crops nor cover crop mixes impact crop yields more than single species cover crops.

Legume cover crops can particularly increase crop yields in degraded, low organic matter, and coarse-textured soils if they produce significant amounts of biomass. Crop yield response to cover crops depends on how cover crops are selected and managed. Overall, non-legume cover crops can maintain or reduce crop yields, but legume cover crops can increase yields under zero or low rates of inorganic N application.

References

Abdalla, M., Hastings, A., Cheng, K., Yue, Q., Chadwick, D., Espenberg, M., Truu, J., Rees, R. M., & Smith, P. (2019). A critical review of the impacts of cover crops on nitrogen leaching, net greenhouse gas balance and crop productivity. *Global Change Biology*, *25*, 2530–2543.

Acharya, P., Ghimire, R., Cho, Y., Thapa, V. R., & Sainju, U. M. (2022). Soil profile carbon, nitrogen, and crop yields affected by cover crops in semiarid regions. *Nutrient Cycling in Agroecosystems*, *122*, 191–203.

Balkcom, K. S., & Reeves, D. W. (2005). Sunn-hemp utilized as a legume cover crop for corn production. *Agronomy Journal*, *97*, 26–31.

Bennett, H. H. (1939). *Soil conservation* (993 p). McGraw-Hill Book Co., Inc.

Blanco-Canqui, H., Claassen, M. M., & Presley, D. R. (2012). Summer cover crops fix nitrogen, increase crop yield and improve soil-crop relationships. *Agronomy Journal, 104*, 137–147.

Blanco-Canqui, H., Ruis, S., Holman, H., Creech, C., & Obour, A. (2022). Can cover crops improve soil ecosystem services in water-limited environments? A review. *Soil Science Society of America Journal, 86*, 1–18.

Blanco-Canqui, H., Shaver, T. M., Lindquist, J. L., Shapiro, C. A., Elmore, R. W., Francis, C. A., & Hergert, G. W. (2015). Cover crops and ecosystem services: Insights from studies in temperate soils. *Agronomy Journal, 107*, 2449–2474.

Brooker, A. P., Renner, K. A., & Sprague, C. L. (2020). Interseeding cover crops in corn. *Agronomy Journal, 112*, 139–147.

Chapagain, T., Lee, E. A., & Raizada, M. N. (2020). The potential of multi-species mixtures to diversify cover crop benefits. *Sustainability, 12*, 2058.

Curran, W. S., Hoover, R. J., Roth, G. W., Wallace, J. M., Dempsey, M. A., Mirsky, S. B., Ackroyd, V. J., Ryan, M. R., & Pelzer, C. J. (2018). Evaluation of cover crops drill-interseeded into corn across the mid-Atlantic. *Crops Soils, 51*, 18.

Deines, J. M., Deines, K., Guan, B., Lopez, Q., Zhou, C. S., White, S., & Wang, D. B. (2023). Recent cover crop adoption is associated with small maize and soybean yield losses in the United States. *Global Change Biology, 29*, 794–807.

Florence, A. M., & McGuire, A. M. (2020). Do diverse cover crop mixtures perform better than monocultures? A systematic review. *Agronomy Journal, 112*, 3513–3534.

Holman, J. D., Arnet, K., Dille, J., Maxwell, S., Obour, A., Roberts, T., Roozeboom, K., & Schlegel, A. (2018). Can cover or forage crops replace fallow in the semiarid Central Great Plains? *Crop Science, 58*, 932–944.

Koehler-Cole, K., Elmore, R. W., Blanco-Canqui, H., Francis, C. A., Shapiro, C. A., Proctor, C. A., Heeren, D. M., Ruis, S., Irmak, S., & Ferguson, R. B. (2020). Cover crop productivity and subsequent soybean yield in the western Corn Belt. *Agronomy Journal, 112*, 2649–2663.

Marcillo, G. S., & Miguez, F. E. (2017). Corn yield response to winter cover crops: An updated meta-analysis. *Journal of Soil and Water Conservation, 72*, 226–239.

Nielsen, D. C., Lyon, D. J., Hergert, G. W., Higgins, R. K., & Holman, J. D. (2015). Cover crop biomass production and water use in the Central Great Plains. *Agronomy Journal, 107*, 2047–2058.

Noland, R. L., Wells, M. S., Sheaffer, C. C., Baker, J. M., Martinson, K. L., & Coulter, J. A. (2018). Establishment and function of cover crops interseeded into corn. *Crop Science, 58*, 863–873.

Ruis, S. J., Blanco-Canqui, H., Creech, C. F., Koehler-Cole, K., Elmore, R. W., & Francis, C. A. (2019). Cover crop biomass production in temperate agroecozones. *Agronomy Journal, 111*, 1535–1551.

Ruis, S. J., Blanco-Canqui, H., Elmore, K. R. W., Proctor, C., Koehler-Cole, K., Shapiro, C. A., Francis, C. A., & Ferguson, R. B. (2020). Impacts of cover crop planting dates on soils after four years. *Agronomy Journal, 112*, 1649–1665.

Ruis, S. J., Blanco-Canqui, H., Jasa, P. J., Ferguson, R. B., & Slater, G. (2017). Can cover crop use allow increased levels of corn residue removal for biofuel in irrigated and rainfed systems? *Bioenergy Research, 10*, 992–1004.

Schipanski, M. E., Barbercheck, M., Douglas, M. R., Finney, D. M., Haider, K., Kaye, J. P., Kemanian, A. R., Mortensen, D. A., Ryan, M. R., Tooker, J., & White, C. (2014). A framework for evaluating ecosystem services provided by cover crops in agroecosystems. *Agricultural Systems, 125*, 12–22.

Sigdel, S., Chatterjee, A., Berti, M., Wick, A., & Gasch, C. (2021). Interseeding cover crops in sugar beet. *Field Crops Research, 263*, 108079.

Stanton, V. L., & Haramoto, E. R. (2021). Biomass potential of drill interseeded cover crops in corn in Kentucky. *Agronomy Journal, 113*, 1238–1247.

Tamburini, G., Bommarco, R., Wanger, T. C., Kremen, C., van der Heijden, M. G. A., Liebman, M., & Hallin, S. (2020). Agricultural diversification promotes multiple ecosystem services without compromising yield. *Science Advances, 6*, eaba1715.

Tonitto, C., David, M. B., & Drinkwater, L. E. (2006). Replacing bare fallows with cover crops in fertilizer-intensive cropping systems: A meta-analysis of crop yield and N dynamics. *Agriculture, Ecosystems and Environment, 112*, 58–72.

13

Grazing and Harvesting

13.1 Cover Crop Biomass Removal

Cover crops were not initially designed for grazing nor harvesting. Their main function was to cover and protect the soil from erosion, maintain or improve soil fertility and productivity, and provide other services related to soil. However, interest is now growing in using cover crops as a component of integrated crop-livestock systems (Figure 13.1). It is considered that reintegrating crops with livestock via grazing crop residues and cover crops can reduce soil degradation, improve environmental quality, and increase overall farm profitability relative to systems with separation of crops from livestock production (Rakkar et al., 2017; Smart et al., 2020). Increasing conversion of grasslands to croplands coupled with increased extreme weather events has reduced forage availability, prompting the need to locate additional forage supplies.

Many consider grazing or harvesting cover crops to be an important piece to support livestock production in seasons when forage availability is limited while diversifying current cropping systems and potentially maintaining or improving soil ecosystem services (Sulc & Franzluebbers, 2014). Grazing cover crops could extend the grazing season and alleviate the sole reliance on crop residues and grasslands or pasturelands for forage supply (Figure 13.1; Drewnoski et al., 2018). Also, unlike crop residues (e.g., wheat straw, corn stover), legume cover crops and their mixes can provide high quality forage (Drewnoski et al., 2018; Franzluebbers & Stuedemann, 2008).

However, concerns revolve around the potential increase in soil compaction from grazing cover crops and reduction in subsequent crop yields. Thus, the questions are: Does cover crop grazing or harvesting adversely affect soil ecosystem services? Does cover crop grazing or harvesting cover crops erase the benefits of non-grazed or non-harvested cover crops? (Drewnoski et al., 2018). The extent of

---

Cover Crops and Soil Ecosystem Services, First Edition. Humberto Blanco.
© 2023 American Society of Agronomy, Inc./Crop Science Society of America, Inc.
/ Soil Science Society of America, Inc. Published 2023 by John Wiley & Sons, Inc.

Figure 13.1 Grazing cover crops can support livestock production while maintaining or improving soil ecosystem services. Photo by M. Drewnoski.

grazing impacts could be influenced by cover crop biomass production, stocking rate or density, length of grazing period, soil conditions during grazing, and others. Our knowledge of the impacts of livestock grazing of pastures or grasslands on soils and biomass production is abundant (Sulc & Franzluebbers, 2014), but how biomass removal from cover crops affects the intended original use of cover crops deserves further discussion. An improved understanding of cover crop grazing impacts can be valuable for management decisions in a time when emphasis on crop integration with livestock is gaining interest. This chapter discusses the magnitude to which grazing and harvesting of cover crops affect all soil ecosystem services based on published data.

13.2 Grazing

13.2.1 Soil Compaction

One of the main concerns about grazing cover crops is the potential increase in soil compaction (Figure 13.2). It is well known that an increase in soil compaction can reduce soil macroporosity, water infiltration, root development, and thus crop production. The pressure exerted by a cow, for instance, is approximately 0.1 MPa when standing and approximately 0.5 MPa when walking (Schomberg et al., 2021). The pressure from a cow on the soil is approximately two-fold greater than exerted

Figure 13.2 Grazing cover crops can cause pugging and compaction especially when the soil is wet. Photo by H. Blanco.

by a farm vehicle. This comparison may justify the reluctance to grazing cover crops, but what does field research really indicate about cover crop grazing impacts on soil compaction?

Soil penetration resistance and bulk density are indicators of soil compaction. Soil penetration resistance measures soil resistance against root penetration while soil bulk density measures how densely packed a soil is. Research shows grazing cover crops does not generally increase soil penetration resistance, indicating grazing cover crops has minimal impacts on soil compaction (Table 13.1). However, wet soil and dry soil can differently respond to compactive forces. Grazing cover crops could increase soil compaction when the soil is wet during the grazing period (Figure 13.2). Data from studies in Georgia indicated that grazing cover crops with cattle increased penetration resistance, but the increase in penetration resistance was confined to the upper 20 cm soil depth and was short-lived when 225–300 kg heifers (35–40 cows) grazed cover crops planted in 1.29 or 1.40 ha catchments (Franzluebbers & Stuedemann, 2008; Schomberg et al., 2021).

Furthermore, when cover crop grazing increases soil penetration resistance, such increase is normally below the threshold level that can significantly reduce root growth (<2 MPa or $20\,\text{kg}\,\text{cm}^{-2}$). This suggests while grazing cover crops can statistically increase soil penetration resistance in some cases, the increase can be of little agronomic significance. Similarly, grazing cover crops can have minimal impacts on soil bulk density. An on-farm study in the central U.S. Great Plains found grazing cover crops with cattle at 5.9 animal unit month (AUM) per hectare over a four-month period in winter and/or spring had no significant effect on soil

Table 13.1 Case Studies Reporting Impacts of Grazing Cover Crops on Soil Penetration Resistance Under Different Locations, Cattle Stocking Rates, and Soil Depths

Location	Soil texture	Years	Cover crop	Cattle stocking rate	Soil depth (cm)	Grazing impact on penetration resistance
[a]South Dakota, U.S.	Loam, clay loam, and silty clay loam	3	Oat	na	0–5	ns
[b]Nebraska, U.S.	Silt loam	5	Oat	1.4–4.0 steer ha^{-1} per 70 d	0–5 and 5–10	ns
[c]Nebraska, U.S.	Sandy loam	3	Rye	5.9 AUM ha^{-1} over 4 mo in winter and/or spring	0–10 and 10–20	ns
[d]Missouri, U.S.	Silt loam	6	Cereal rye and winter wheat	7.2–13.2 AUM ha^{-1} (grazed 2 to 4 times in spring)	0–5	Increased: 0.7–0.9 MPa
					5–10	Increased: 1.1–1.6 MPa
					10–15	Increased: 1.2–1.7 MPa
					15–20	Increased: 1.4–1.7 MPa
					20–35	ns
[e]Georgia, U.S.	Sandy loam and sandy clay loam	4	Rye	225–300 kg heifers (35–40 cows) 1.29 or 1.40 ha catchments	0–2.5	ns
					2.5–7.5	Increased: 1.4–2.0 MPa
					7.5–15	Increased: 2.2–2.6 MPa
					15–30	ns

[a] Dhaliwal and Kumar (2021).
[b] Anderson et al. (2022).
[c] Blanco-Canqui, Drewnoski, et al. (2020).
[d] Dhakal et al. (2022).
[e] Schomberg et al. (2021).
na means not available, while ns means no significant impact; AUM, animal unit month.

bulk density during a three-year study (Blanco-Canqui, Drewnoski, et al., 2020). Further, a two-year on-farm study in the central U.S. Great Plains across various locations found grazing cover crops did not alter soil bulk density relative to no cover crops, which suggests grazing cover crops does not compact soil (Kelly et al., 2021).

The following mechanisms can explain the limited impacts of grazing cover crops on soil compaction. One, organic matter return from livestock in the form of manure can improve soil elasticity and rebounding capacity against compressive forces. Research shows that animal manure-derived organic matter reduces the soil's susceptibility to compaction and compressibility (Soane, 1990; Anderson et al., 2022). Two, animal manure can enhance soil aggregation, macroporosity, and microbial activity, which can improve soil resistance to compaction. Three, freezing–thawing and drying-wetting cycles, which are common in temperate regions, can alleviate any grazing-induced increases in soil compaction (Anderson et al., 2022). Overall, grazing cover crops minimally affects soil compaction although grazing wet soils can increase soil compaction risks.

13.2.2 Water Infiltration

While it is thought that grazing cover crops can reduce macroporosity and thus reduce water infiltration, the few published studies show grazing cover crops does not have large negative effects on water infiltration relative to non-grazed cover crops (Table 13.2). Two case studies found no effect of grazing cover crops on water infiltration, while one study found that grazing winter rye cover crop at a high cattle stocking rate reduced water infiltration in one out of three years (Table 13.2). If grazing were to reduce soil porosity, particularly macroporosity,

Table 13.2 A Few Case Studies on the Impacts of Grazing Cover Crops on Water Infiltration

Location	Years	Cover crop	Cattle stocking rate	Grazing impact on water infiltration
[a]South Dakota, U.S.	3	Oat	na	ns
[b]Nebraska, U.S.	5	Oat	1.4–4.0 steer ha^{-1} per 70 days	ns
[c]Nebraska, U.S.	3	Rye	5.9 AUM ha^{-1} over 4 mo in winter and/or spring	No effect in 2 of 3 yr Reduced in 1 of 3 yr

[a] Dhaliwal and Kumar (2021).
[b] Anderson et al. (2022).
[c] Blanco-Canqui, Drewnoski, et al. (2020).
na means not available, while ns means no significant impact; AUM, animal unit month.

then it would rapidly reduce water infiltration. However, the lack of significant changes in soil bulk density with grazing cover crop grazing suggests that grazing cover crops may not significantly alter soil porosity (Blanco-Canqui, Drewnoski, et al., 2020; Kelly et al., 2021). Soil aggregation is another mechanism that affects water infiltration rate. Research indicates grazing cover crops does not commonly reduce soil aggregate stability compared with non-grazed cover crops (Blanco-Canqui, Drewnoski, et al., 2020; Kelly et al., 2021) The limited or no effect of grazing cover crops on water infiltration and related soil properties indicates grazing may not erase the benefits of cover crops for capturing precipitation water.

13.2.3 Soil Carbon Dynamics and Sequestration

Cover crops generally accumulate C in the soil especially in the long term if cover crop biomass production is sufficiently high (>1 Mg ha^{-1}) on a consistent basis and initial soil C level is low. Thus, one might expect that grazing cover crops can erase or reduce the benefits of cover crops for sequestering C because it directly removes C-enriched biomass. However, available data indicate grazing cover crops does not adversely affect the ability of cover crops to accumulate C in the soil (Table 13.3). Also, grazing cover crops may not increase the rate of CO_2

Table 13.3 Case Studies on The Impacts of Grazing Cover Crops on Soil Organic C Concentration Under Different Locations and Cattle Stocking for the Upper 30 cm Soil Depth

Location	Years	Cover crop	Cattle stocking rate	Grazing impact on soil C
[a]South Dakota, U.S.	3	Oat	na	ns
[b]Nebraska, U.S.	5	Oat	1.4–4.0 steer ha^{-1} per 70 days	ns
[c]Kansas, Nebraska, and Colorado, U.S.	2	Mix	307–1052 kg live weight ha^{-1} for 28 days	ns
[d]Kansas, U.S.	6	Oat and triticale	877–1755 kg live weight ha^{-1}	ns
[e]Nebraska, U.S.	3	Rye	5.9 AUM ha^{-1} over 4 mo in winter and/or spring	ns

[a] Dhaliwal & Kumar (2021).
[b] Anderson et al. (2022).
[c] Kelly et al. (2021).
[d] Simon et al. (2021).
[e] Blanco-Canqui, Drewnoski, et al. (2020).
na means not available, while ns means no significant impact; AUM, animal unit month.

emissions. At two sites in South Dakota, grazing cover crops with cattle, which removed no more than one-half of cover crop biomass, had no effect on CO_2 and N_2O fluxes nor soil organic C concentration relative to non-grazed cover crops during a two-year study (Singh et al., 2020). Thus, research data indicate grazing cover crops with cattle may not reduce soil C gains from cover crops.

It is important to discuss that the addition of C-enriched manure from grazing animals could negate the adverse impacts of biomass removal on soil C sequestration. Also, grazing cover crops does not remove the belowground biomass (roots). As discussed in an earlier chapter, plant roots can be more essential than aboveground biomass for maintaining or increasing soil C concentration (Xu et al., 2021). The magnitude to which grazing cover crops impacts the soil C cycle in the long term is still unclear as most of the available studies are short term (<5 years). Soil organic C concentration often changes slowly after management shift. Thus, it is expected that grazing cover crop impacts on soil C may be more measurable in the long term than in the short term. At this point, however, available data indicate that grazing cover crops does not diminish the potential of cover crops to sequester C in the soil.

13.2.4 Crop Yields

Research indicates grazing cover crops does not generally reduce subsequent crop yields. The summary of recent studies in Table 13.4 indicates crop yields are unaffected by cover crop grazing relative to non-grazed cover crops in most cases. Crop yields will likely not decrease following cover crop grazing unless there are large changes in soil properties, such as excessive compaction (penetration resistance >2 MPa). For example, on a claypan soil in Missouri, grazing cover crops with cattle at high stocking rates (7.2–13.2 AUM ha^{-1}) increased soil penetration resistance to 20 cm depth and reduced corn yield by 0.4 Mg ha^{-1} but not soybean yield (Dhakal et al., 2022). The study found grazing cover crops increased penetration resistance to values close to 2 MPa in some years for the 5–20 cm depth. Thus, grazing cover crop effects on crop yields can be highly site-specific, depending on stocking rates and soil conditions during grazing.

The small or no effects of grazing impact on subsequent crop yields in most cases (Table 13.4) lead to the question: Why doesn't grazing cover crops reduce yields? The limited grazing impacts on crop yields are attributed to the limited impacts of grazing on soil ecosystem services that affect crop production. In general, grazing cover crops does not compact soil above the threshold levels nor reduces organic matter concentration, soil water, and weed suppression potential relative to non-grazed cover crops. First, grazing animals can offset nutrient removal with biomass and maintain soil fertility by returning manure. Manure return also activates microbial activity, which contribute to nutrient cycling and

Table 13.4 Crop Yield Response to Cover Crop Grazing Under Different Locations and Stocking Rates

Location	Cover crop	Cattle stocking rate	Years	Grazing impact on crop yield
[a]Missouri, U.S.	Wheat, rye, and ryegrass	7.2–13.2 AUM ha^{-1} (grazed 2–4 times in spring)	6	Reduced corn yield by 0.4 Mg ha^{-1}; No effect on soybean yield
[b]Nebraska, U.S.	Oat	1.4–4.0 steer ha^{-1} per 70 d	5	ns
[c]Western Cape, South Africa	Mix	9975 kg live sheep weight ha^{-1} for 10 d	2	ns
[d]Kansas, Nebraska, and Colorado, U.S.	Mix	307–1052 kg live weight ha^{-1} for 28 d	2	ns
[e]South Dakota, U.S.	Mix	7.7 AUM ha^{-1} for 11 d	4	ns

[a] Dhakal et al. (2022).
[b] Anderson et al. (2022).
[c] Smit et al. (2021).
[d] Kelly et al. (2021).
[e] Rai et al. (2021).
ns means no significant impact; AUM, animal unit month.

soil fertility maintenance (Franzluebbers & Stuedemann, 2007). Second, grazed crops could still suppress weeds if biomass left is high enough such as under low or moderate grazing (MacLaren et al., 2018). Also, livestock can also consume weeds during grazing. Third, grazing cover crops does not often reduce soil water for the next crop more than non-grazed cover crops (Kelly et al., 2021). In the long term, accumulation of organic matter with manure could increase soil ability to retain water more than in systems without manure input although the amount of manure added has not been quantified in the available studies. In general, data indicate that decline in crop yields is not expected in most cases under current cover crop grazing scenarios (Table 13.4).

13.3 Minimizing Potential Grazing Impacts

While the discussion above indicates small or no negative effects of grazing cover crops on soil ecosystem services, cover crop and livestock management is critical to minimize any potential negative effects of grazing cover crops on soil ecosystem services. Among the site-specific factors that can affect grazing impacts on soil

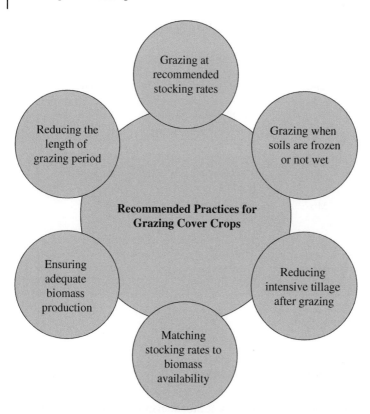

Figure 13.3 Strategies for alleviating grazing cover crop impacts on soil compaction and other soil ecosystem services.

properties and crop yields include amount of cover crop biomass removal, stocking rate, years under grazing, and soil water content (Figure 13.3).

13.3.1 Amount of Biomass Removal

While one may consider that the amount of cover crop biomass removed by livestock could affect the extent of grazing impacts on soil ecosystem services, research data do not entirely support this consideration. Most cover crop grazing studies found that even high amounts of cover crop biomass removal (>90%) by grazing can have small or no effects on crop yields and other soil services. In a humid subtropical region, cattle removed approximately 94% of pearl millet and winter rye cover crop biomass in a four-year study, but differences in soil compaction and crop yields between grazed and non-grazed cover crops were small (Franzluebbers & Stuedemann, 2008). Also, in the central Great Plains, Blanco-Canqui, Drewnoski, et al. (2020) reported that grazing cattle removed approximately 92% of winter rye cover crop biomass (8.6 ± 3.3 Mg ha^{-1}); yet it had no major negative effects on soils and crop yields after three years. These short-term studies (<4 years) indicate that even when significant amounts of cover crop biomass are removed, grazing effects on soil properties and crop yields can still be marginal.

13.3.2 Stocking Rate

Animal stocking rate, which refers to the number of animals per unit of cropland, not only determines the amount of biomass removed during grazing but also the pressure exerted in the soil that causes compaction. It is well recognized that one single pass of a farm tractor can cause approximately 70% of soil compaction. Thus, it is expected that a short-term high stocking rate can rapidly compact soil as soon as animals enter the field, especially when the soil is wet. Available studies indicate grazing cover crops at stocking rates below the recommended rates does not generally increase soil compaction above the root growth threshold levels. The recommended stocking rate may vary between 4 and 6 AUM ha^{-1}, depending on cover crop biomass production (Anderson et al., 2022).

Studies on cover crop grazing under different stocking rates within the same experiment are unavailable to establish the recommended stocking rates. In the western U.S. Corn Belt, corn residue grazing in spring at cattle stocking rates as high as 13 AUM ha^{-1} did not reduce subsequent corn and soybean yields even when causing some compaction during a 16-year study (Rakkar et al., 2017). Also, across three years, cattle stocking rate at 5.9 AUM ha^{-1} for approximately four months in winter and spring reduced winter rye cover crop biomass by 92% but had minimal effects on soil compaction and crop yields (Blanco-Canqui, Drewnoski, et al., 2020). High stocking rates could cause compaction and negatively impact crop yields in some years. Matching stocking rate to cover crop biomass availability is important to minimize effects on soil compaction (Schomberg et al., 2021; Figure 13.3).

13.3.3 Years under Grazing

One may hypothesize that repeated cover crop grazing events year after year may have cumulative effects on soil ecosystem services. However, research data indicate that the impacts of grazing cover crops for multiple years do not seem to be cumulative although long-term studies specifically monitoring cover crop grazing impacts with time are limited. A corn residue grazing study in the western Corn Belt under no-till corn-soybean rotation found that residue grazing had few effects after 16 years of grazing at high stocking rates (9.3–13 AUM ha^{-1}; Rakkar et al., 2017). Year-to-year variability in soil compaction due to cover crop grazing appears to be greater than any cumulative effects of grazing with time. Grazing cover crops increases penetration resistance in some years but not in others with limited cumulative effects (Blanco-Canqui, Drewnoski, et al., 2020; Franzluebbers & Stuedemann, 2008). Fluctuations in weather, freezing and thawing cycles, and increased organic matter with manure could alleviate any potential cumulative impacts of grazing.

13.3.4 Tillage

Tillage is sometimes used to correct soil compaction and surface roughness induced by grazing animals (Clark et al., 2004). Tillage can temporarily break compacted layers and alleviate compaction. However, tilled soils can rapidly reconsolidate after tillage and become more susceptible to compaction than no-till soils in the long term due to lower soil organic C and less favorable soil physical properties. The increase in soil organic matter concentration in no-till systems near the soil surface can reduce bulk density, and promote soil aggregation and formation of biopores (e.g., earthworm and root channels), which can enhance soil resistance to compaction. Franzluebbers and Stuedemann (2008) found grazing cover crops tended to increase soil penetration resistance in conventionally tilled soils but had no effect in no-till soils compared with non-grazed cover crops across 10 sampling events.

Further, because tillage accelerates decomposition of organic matter and C loss potential, it could reduce the soil C benefits of manure from grazing animals. Frequent tillage to correct soil compaction and surface roughness induced by grazing may negatively affect other soil properties in the long term although reduced till or one-time tillage may have limited adverse impacts. When grazing cover crops increases soil compaction above the threshold levels of root growth, the use of reduced till such as strip tillage can be an option before planting crops to alleviate the potential adverse effects of soil compaction on subsequent crop yields. A study in the central U.S. Great Plains found no differences in corn silage yield between grazed and non-grazed cover crops after three years under high cattle stocking rates (5.9 AUM) for approximately four months in winter and spring when strip till was used before planting corn (Blanco-Canqui, Drewnoski, et al., 2020). Thus, the use of reduced till such as strip till following cover crop grazing can alleviate compaction effects on crop yields. The use of plow till or intensive tillage to alleviate soil compaction can degrade soil properties more than reduced till and it thus should be discouraged.

13.3.5 Soil Water Content

Soil water content at the time of cover crop grazing is one of the most critical determinants of soil compaction severity. Soil compactibility (soil's susceptibility to compaction) is highly and positively correlated with an increase in soil water content up to a critical water content, which is the soil water content at which maximum compactibility occurs. Under identical animal stocking rates, wet soils are more compactable than dry soils because water molecules lubricate soil particles, facilitate soil particle sliding, and promote soil packing relative to dry soils. In the southeastern U.S., Schomberg et al. (2021) reported grazing cover crops increased near-surface soil penetration resistance only when the soil was wet during spring.

Grazing cover crops when soil is frozen or when soil is relatively dry can be a strategy to reduce soil compaction risks (Figure 13.3). A study in the U.S. Midwest found that grazing corn stover in a corn-soybean system increased soil penetration resistance and reduced subsequent soybean yield but grazing did not reduce soybean yields if grazing was confined to periods when soil temperature was below 0 °C (Clark et al., 2004). The effect of soil water content on soil compaction can be larger in clayey soils than in soils with low clay content (Dhakal et al., 2022). Restricting the presence of livestock only to periods when soil is frozen can reduce concerns of soil compaction from grazing cover crops.

13.4 Harvesting

Harvesting cover crops for livestock or biofuel production can be another expanded use from cover crops (Figure 13.4; Krueger et al., 2011). Crop residues

Figure 13.4 Winter rye cover crop planted after corn silage and harvested as forage in the eastern U.S. Great Plains. Photo by H. Blanco.

such as corn stover are often harvested to support livestock production but cover crops are also harvested in a few cases to provide forage and complement crop residues. Harvested fresh cover crops such as from legume and non-legume mixes can be more palatable and richer in nutrients than grass crop residues (Drewnoski et al., 2018). The question is: Can cover crops be harvested without adverse effects on soil ecosystem services? This method of cover crop biomass removal could have different impacts on soil services from grazing.

Forage harvester traffic can simulate livestock traffic during grazing but harvesting cover crops does not return animal manure unlike grazing. As a result, one may hypothesize that harvesting cover crops could have more adverse effects on soil ecosystem services than grazing in the long term. Studies on cover crop harvesting or haying are emerging (Holman et al., 2018). A review found cutting cover crops at 7.5 to 10 cm of stubble height does not generally affect soil properties, soil fertility, weed suppression, and crop yields (Blanco-Canqui, Ruis, et al., 2020).

Similar to grazing cover crops, effects of harvesting cover crops should be assessed on a management-specific basis as harvesting effects can vary with cover crop management such as harvest timing. A study in Minnesota reported harvesting winter rye cover crop at 7–17 cm stubble height under corn silage systems reduced soil water content, available N, and corn silage yield compared with non-harvested and no cover crop during a two-year study when winter rye was harvested only two days before corn planting (Krueger et al., 2011). However, a three-year study in Nebraska found harvesting a cover crop mix of legumes and non-legumes in fall (approximately five months before corn planting) had no effect on soil penetration resistance and bulk density, wet aggregate stability, and nitrate leaching potential (Blanco-Canqui et al., 2020). The same study found that cover crop harvesting did not reduce corn silage yields in two of three years, but reduced cumulative infiltration and organic matter concentration compared with non-harvested cover crops (Blanco-Canqui, Drewnoski, et al., 2020). These few studies indicate harvest timing can play a major role in harvesting effects.

Research suggests proper harvest timing and cutting heights may not eliminate all the benefits of cover crops (Blanco-Canqui, Ruis, et al., 2020). High cutting heights, which mimic low or moderate grazing, can leave sufficient cover crop standing and residues to maintain or improve soil services (Holman et al., 2018). Similar to grazing, harvesting cover crops does not remove roots, which are essential to soil services from cover crops. Research also indicates the harvestable cover crop biomass amount without adversely soil ecosystem services can range from 1 to 3 Mg ha^{-1} in semiarid and from 1 to 6 Mg ha^{-1} in humid regions (Blanco-Canqui, Ruis, et al., 2020). In sum, harvesting cover crops can be a promising management strategy to obtain an additional service (forage) from cover crops without significantly compromising other services from cover crops under proper cover crop termination timing and stubble cutting heights.

13.5 Grazing and Harvesting: An Added Benefit from Cover Crops?

Based on the available research information, grazing and harvesting cover crops are highly promising and should not be viewed as a negative practice that degrade soil ecosystem services. Data suggest cover crops can deliver an additional ecosystem service, which is forage for livestock production or feedstock for cellulosic biofuel production without major negative consequences on other soil ecosystem services. This can defy the original concept for cover crop adoption. Note grazing or harvesting cover crops leaves intact all the belowground biomass (roots) and a portion of aboveground biomass. Thus, grazing or harvesting of cover crops may not undermine the soil benefits of non-grazed cover crops as it only removes a portion of cover crop residues. The key to maintain the intended soil conservation benefits of cover crops will be how cover crops are grazed or harvested (Figure 13.3).

The limited adverse effects of grazing cover crops on soil ecosystem services mirror those of crop residue grazing (Rakkar et al., 2017). Both cover crops and crop residues can be grazed to support livestock production while maintaining their benefits for soil ecosystem services. Furthermore, while water use by cover crops is a limiting factor for cover crop adoption in water-limited regions, grazing cover crops may limit water use by maintaining biomass levels to low or optimum levels while still protecting soil and delivering essential ecosystem services (Kelly et al., 2021). In high precipitation regions, grazed cover crops could maintain soil and water management benefits from cover crops as differences between grazed and non-grazed cover crops are small or negligible.

Research findings support the idea that adding cover crops to current cropping systems can be an option to bring animals back into croplands under proper stocking rates and soil conditions. Cover crops and livestock can complement each other within the reemerging crop–livestock integrated systems while maintaining soil ecosystem services. Cover crops can provide high quality forage whereas grazing livestock can provide nutrient-enriched manure for enhanced crop and cover crop production, leading to a positive mutual relationship between cover crop and livestock production. Also, grazing animals remove biomass C but return C with manure, which can offset soil C losses from grazed fields. While the few data do not show large negative impacts of grazing cover crops on soil ecosystem services, additional long-term data can be helpful to make definitive conclusions about grazing cover crop impacts.

13.6 Summary

Available data indicate that livestock presence in fields with cover crops has minimal impacts on soil compaction, soil C, crop yields, and other soil ecosystem services. Cover crop and livestock management can influence the magnitude of cover crop grazing impacts. Grazing cover crops when the soil is frozen or when soil is relatively dry can be a viable approach to avoid excessive soil compaction. High animal stocking rates can cause compaction but grazing at recommended stocking rates should not have major negative ramifications for soil ecosystem services.

Grazing increases soil penetration resistance in some cases. However, such increase is often not high enough to limit root growth and reduce crop yields. Manure deposition by the grazing animals, remaining aboveground cover crop residues, and cover crop roots most likely maintain soil properties and reduce the soil's susceptibility to compaction, mitigating any negative impacts of grazing on crop yields and other ecosystem services (e.g., C sequestration). Frequent freezing–thawing cycles such as in cool temperate regions can also contribute to compaction mitigation.

Managing animal stocking rate by synchronizing stocking rate with the amount of cover crop biomass available can be critical to minimize any adverse effects on soils and crops. Similar to grazing, harvesting cover crops at high stubble cutting heights (>7.5 cm) does not necessarily erase or reduce the soil and crop benefits of cover crops. Overall, data available to this point indicate cover crops can be used as a temporary pasture to augment forage supply and contribute to crop-livestock integration without damaging soils or reducing crop yields.

References

Anderson, L. H., Blanco-Canqui, M. D., & MacDonald, J. (2022). Cover crop grazing impacts on soil properties and crop yields under irrigated no-till corn-soybean. *Soil Science Society of America Journal, 86*, 118–133.

Blanco-Canqui, H., Drewnoski, M., Redfearn, D., Parsons, J., Lesoing, G., & Tyler, W. (2020). Does cover crop grazing damage soils and reduce crop yields? *Agrosystems Geosciences Environment, 3*, e20102.

Blanco-Canqui, H., Ruis, S., Proctor, C., Creech, C., Drewnoski, M., & Redfearn, D. (2020). Harvesting cover crops for biofuel and livestock production: Another ecosystem service? *Agronomy Journal, 112*, 2373–2400.

Clark, J. T., Russell, J. R., Karlen, D. L., Singleton, P. L., Busby, W. D., & Peterson, B. C. (2004). Soil surface property and soybean yield response to corn Stover grazing. *Agronomy Journal, 96*, 1364–1371.

Dhakal, D., Erwin, Z. L., & Nelson, K. A. (2022). Grazing cover crops in a no-till corn and soybean rotation. *Agronomy Journal, 114*, 1255–1268.

Dhaliwal, J. K., & Kumar, S. (2021). Hydro-physical soil properties as influenced by short and long-term integrated crop–livestock agroecosystems. *Soil Science Society of America Journal, 85*, 789–799.

Drewnoski, M., Parsons, J., Blanco-Canqui, H., Redfearn, D., Hales, K., & McDonald, J. (2018). Forages and pastures symposium: Cover crops in livestock production: Whole-system approach. Can cover crops pull double duty: Conservation and profitable forage production in the midwestern United States. *Journal of Animal Science, 96*, 3503–3512.

Franzluebbers, A. J., & Stuedemann, J. A. (2007). Crop and cattle responses to tillage systems for integrated crop–livestock production in the southern Piedmont, USA. *Renewable Agriculture and Food Systems, 22*, 168–180.

Franzluebbers, A. J., & Stuedemann, J. A. (2008). Soil physical responses to cattle grazing cover crops under conventional and no tillage in the southern Piedmont USA. *Soil and Tillage Research, 100*, 141–153.

Holman, J. D., Arnet, K., Dille, J., Maxwell, S., Obour, A., Roberts, T., Roozeboom, K., & Schlegel, A. (2018). Can cover or forage crops replace fallow in the semiarid central Great Plains? *Crop Science, 58*, 932–944.

Kelly, C., Schipanski, M. E., Tucker, A., Trujillo, W., Holman, J. D., Obour, A. K., Johnson, S. K., Brummer, J. E., Haag, L., & Fonte, S. J. (2021). Dryland cover crop soil health benefits are maintained with grazing in the U.S. High and Central Plains. *Agriculture, Ecosystems and Environment, 313*, 107358.

Krueger, E. S., Ochsner, T. E., Porter, P. M., & Baker, J. M. (2011). Winter rye cover crop management influences on soil water, soil nitrate, and corn development. *Agronomy Journal, 103*, 316–323.

MacLaren, C., Storkey, J., Strauss, J., Swanepoel, P., & Dehnen-Schmutz, K. (2018). Livestock in diverse cropping systems improve weed management and sustain yields whilst reducing inputs. *Journal of Applied Ecology, 56*, 144–156.

Rai, T. S., Nleya, T., Kumar, S., Sexton, P., Wang, T., & Fan, Y. (2021). The medium-term impacts of integrated crop–livestock systems on crop yield and economic performance. *Agronomy Journal, 113*, 5207–5221.

Rakkar, M. K., Blanco-Canqui, H., Drijber, R. A., Drewnoski, M. E., MacDonald, J. C., & Klopfenstein, T. (2017). Impacts of cattle grazing of corn residues on soil properties after 16 years. *Soil Science Society of America Journal, 81*, 414–424.

Schomberg, H. H., Endale, D. W., Balkcom, K. S., Raper, R. L., & Seman, D. H. (2021). Grazing winter rye cover crop in a cotton no-till system: Soil strength and runoff. *Agronomy Journal, 113*, 1271–1286.

Simon, L. M., Obour, A. K., Holman, J. D., Johnson, S. K., & Roozeboom, K. L. (2021). Forage accumulation of spring and summer cover crops in western Kansas. *Kansas Agricultural Experiment Station Research Reports, 7* (5). https://doi.org/10.4148/2378-5977.8134

Singh, N., Abagandura, G. O., & Kumar, S. (2020). Short-term grazing of cover crops and maize residue impacts on soil greenhouse gas fluxes in two Mollisols. *Journal of Environmental Quality, 49,* 628–639.

Smart, A. J., Redfearn, D., Mitchell, R., Wang, T., Zilverberg, C., Bauman, J., Derner, J. D., Walker, J., & Wright, C. (2020). Integration of crop-livestock systems: An opportunity to protect grasslands from conversion to cropland in the US Great Plains. *Rangeland Ecology & Management.* https://doi.org/10.1016/j.rama.2019.12.007

Smit, E. H., Strauss, J. A., & Swanepoel, P. A. (2021). Utilisation of cover crops: Implications for conservation agriculture systems in a Mediterranean climate region of South Africa. *Plant and Soil, 462,* 207–218.

Soane, B. D. (1990). The role of organic matter in soil compactibility: A review of some practical aspects. *Soil and Tillage Research, 16,* 179–201.

Sulc, R. M., & Franzluebbers, A. J. (2014). Exploring integrated crop–livestock systems in different ecoregions of the United States. *European Journal of Agronomy, 57,* 21–30.

Xu, H., Vandecasteele, B., De Neve, S., Boeckx, P., & Sleutel, S. (2021). Contribution of above- versus belowground C inputs of maize to soil organic carbon: Conclusions from a 13C/12C-resolved resampling campaign of Belgian croplands after two decades. *Geoderma, 383,* 114727.

14

Economics

14.1 Cover Crops and Farm Profits

Do cover crops increase farm profits? This may be one of the questions that many have before introducing cover crops into a cropping system. Indeed, producers can be reluctant to adopt cover crops if no immediate economic benefit is to be obtained. Thus, economic costs associated with cover crop adoption need discussion. Cover crop adoption has direct and indirect costs and benefits (Bergtold et al., 2017). Some direct costs include seed purchase and planting and termination operations, while indirect costs include reduced water for the next crop in water-limited regions, slow soil warming due to increased cover crop residue cover, and a potential decrease in subsequent crop yields.

Furthermore, addition of cover crops complicates the management of the whole farm system. Cover crop establishment costs can range from $50 to $200 per hectare, depending on seed cost, seeding rates, cover crop species, and planting and termination method (Pratt et al., 2014; Roth et al., 2018). Direct or indirect economic benefits from cover crop adoption are still unclear. The main direct economic benefit would be a potential increase in subsequent crop yields while indirect benefits can include savings on fertilizer, herbicides, and other production costs (Cates et al., 2018).

Literature is replete with information on the overall positive impacts of cover crops on soil physical, chemical, and biological properties; erosion control; reduced nutrient losses; and other ecosystem services related to soils. The follow-up question is: Do these soil benefits result in increased farm income after cover crop introduction relative to not planting cover crops? Planting cover crops in the name of soil conservation and management is a worthy goal, and this is enough reason to further promote cover crop adoption. However, if cover crops do not

Cover Crops and Soil Ecosystem Services, First Edition. Humberto Blanco.
© 2023 American Society of Agronomy, Inc. / Crop Science Society of America, Inc.
/ Soil Science Society of America, Inc. Published 2023 by John Wiley & Sons, Inc.

generate economic returns, then large-scale cover crop adoption may be limited. A direct economic benefit from cover crops can be the most appealing incentive to adopt cover crops.

Interest in cover crops is growing, but their adoption is still slow in spite of the well-known soil benefits of cover crops. In the U.S., the adoption of cover crops does not exceed 5% in most locations (Dunn et al., 2016). Economics of cover crops under different scenarios of cover crop management need further evaluation and discussion for decision-making purposes. In general, it is expected that fields without cover crops will be more profitable than fields with cover crops if costs of cover crop establishment and management are not offset. This chapter highlights some of the opportunities that can generate economic returns from cover crop management.

14.2 Economic Analysis

Published literature has focused more on the implications of cover crop adoption on soil conservation, soil C sequestration, weed suppression, soil fertility, crop yields, and other soil ecosystem services and less on the economics of cover crops. For instance, our understanding of direct and indirect cover crop production costs as well as direct and indirect economic benefits accounting for monetary and nonmonetary soil ecosystem services provided by cover crops through a comprehensive economic analysis is still limited. Some studies have conducted a partial budget economic analysis based on cover crop cost (seed, planting, and termination) and crop yield response (Figure 14.1). In the U.S. Corn Belt, Pratt et al. (2014) estimated that cover crop costs ranged from \$81.76 to \$172.50 ha^{-1}, while economic benefits ranged from \$91.45 to \$192.07 ha^{-1}. Economic benefits can include nutrient scavenging, increased soil organic matter concentration, weed suppression, and reduced nitrate leaching, soil erosion, and soil compaction. Another study in the U.S. Midwest found that a cereal rye and daikon radish cover crop mix can recover approximately 61% of cover crop implementation costs by improving N cycling and reducing soil erosion and N losses to subsurface drainage under corn and soybean systems (Roth et al., 2018). In water-limited regions, Bergtold et al. (2017) estimated that if cover crops replace fallow in crop–fallow systems, positive net returns can be expected.

The few available estimates above suggest cover crops can be profitable, but such profitability varies based on a number of factors including cover crop management, cover crop species, main crop management, and climatic conditions. As an example, the seeding rate for aerial cover crop seeding can be twice of drilled cover crops although aerial seeding does not often outperform drilled cover crops (Blanco-Canqui et al., 2017). Some of the potential opportunities that could

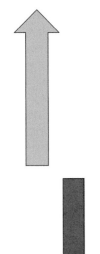

INCOME
1. Saving on forage cost
2. Potential increase in crop yields
3. Fertilizer savings
4. Herbicide savings
5. Soil C credits
6. Biofuel production
7. Soil erosion reduction
8. Savings on non-marketable ecosystem services

EXPENSES
1. Cover crop seed
2. Planting and termination costs
3. Fertilization and irrigation (if any)
4. Grazing (e.g., fences, solar panels, cost of water tank)
5. Potential decrease in crop yield

Figure 14.1 Simplified balance of input and output costs following cover crop introduction.

generate economic income from planting cover crops include grazing and harvesting cover crops, weed suppression, C and N credits, valuation of soil ecosystem services, and others as discussed next.

14.2.1 Grazing and Harvesting

Grazing and harvesting cover crops are emerging opportunities to attain annual positive net return from cover crops. Grazing or harvesting cover crops does not eliminate all the benefits of cover crops for maintaining or improving the services that soils provide as discussed previously (Tables 13.1–13.4). Cover crop root biomass in addition to the remaining portion of the aboveground biomass coupled with manure input can maintain soil benefits from cover crops when cover crops are grazed (Franzluebbers & Stuedemann, 2007). For instance, in dairy farms, harvested cover crops can be used as a nutritious supplement forage source (Blanco-Canqui et al., 2021). The potential increase in farm economics with grazing and harvesting may offset a decrease in crop yield although excessive soil compaction due to high animal stocking rates during wet periods could increase production costs through the use of tillage operations to alleviate soil compaction.

Studies using a partial budget analysis showed grazing and harvesting cover crops can, in general, increase farm income (Table 14.1). These studies considered positive effects (e.g., reduced forage cost) and negative effects (cover crop seed, cover crop planting and termination costs, fencing) to evaluate cover crop economics. Net farm income from grazing cover crops can increase after the first year due to greater initial investment cost in the first year. For example, in South Dakota, annual net return was $42.56 ha^{-1} in the first year and $107.72 ha^{-1} in the second year after deduction of one-time purchase costs (e.g., fences, solar panel, water tank) from the first year. Table 14.1 also indicates net returns vary from study to study, which suggests that total net economic return depends on cover crop management. It is important to note that the available economic analyses are from short-term studies (<5 years). Multiple years of cover crop management are needed to establish firm economic estimations from cover crop grazing or harvesting.

Amount of biomass, cover crop species, and climate are some of the factors that can affect economic return from grazing and harvesting cover crops. In water-limited regions, grass cover crops can be more profitable than legume cover crops

Table 14.1 Impact of Grazing and Harvesting Cover Crops on Annual Net Return

Location	Cover crop	Years	Method of biomass removal	Annual net return ha^{-1}
[a]South Dakota, U.S.	Grass and legume mixes	2	Grazing	$42.56 in Year 1; $107.72 after Year 1
[b]Georgia, U.S.	Cereal rye	4	Grazing	$81
[c]Georgia, U.S.	Cereal rye and pearl millet	3	Grazing	$302
[d]Alabama, U.S	Oat and ryegrass	3	Grazing	$185–$200
[e]Kansas, U.S	Hairy vetch, winter lentil, spring lentil, winter pea, winter triticale, and spring triticale	5	Harvesting	Spring triticale: 26%; Winter triticale: 240%; Legumes: No net return

[a] Tobin et al. (2020).
[b] Schomberg et al. (2014).
[c] Franzluebbers and Stuedemann (2007).
[d] Siri-Prieto et al. (2007).
[e] Holman et al. (2018).

because grasses cost less (e.g., seed) and produce more biomass than legumes. In a semiarid region, Holman et al. (2018) reported winter and spring triticale cover crops generated positive economic returns and offset the cost of growing cover crops relative to Austrian pea, lentil, and hairy vetch when cover crops were grown during fallow in winter-wheat fallow systems. They also reported that cover crop biomass production was 2.093 Mg ha^{-1} for spring triticale, 4.116 Mg ha^{-1} for winter triticale and 0.695 Mg ha^{-1}, on average, for legumes. The increased cover crop biomass production can thus translate into increased economic benefits when cover crops are harvested as forage.

Interest is increasing in growing cover crop mixes as livestock feed, but research indicates income from cover crop mixes may not be greater than from growing monocultures. Cost of cover crop seeds for mixtures is higher but biomass production is lower or similar to monocultures (Florence & McGuire, 2020). Increasing seeding rate for mixtures could increase their biomass production, but it will increase production costs relative to single species. Also, tillage systems may not influence how grazing and harvesting cover crops affect farm income. Franzluebbers and Stuedemann (2007) reported positive economic return from grazing cover crops can be higher than from non-grazed cover crops regardless of tillage systems.

Precipitation determines the amount of biomass available for removal and thus economic return. If cover crop biomass production is low, then farm profits from fields with cover crops can be minimal or negative. Using cover crops with flexible crop rotations in response to fluctuating precipitation input can be a strategy to generate farm profits in water-limited years. Holman et al. (2018) recommended that growing cover crops as forage in wet years and using fallow in dry years can be more profitable than planting cover crops every year.

Available information suggests generating income from cover crops by grazing or harvesting can be an option to overcome the potential decline in crop yields and increase total positive net returns from the farm. The economic benefits of grazing cover crops can be immediate if cover crops produce sufficient biomass. However, any potential economic benefits from C sequestration and improvement in soil properties can manifest in the long term (>5 years) as it can take time for cover crops to accumulate C and exert changes in soil properties. Available studies show grazing cover crops can at least offset the establishment costs (Bergtold et al., 2017; Holman et al., 2018). Overall, using cover crops as livestock feed or as feedstock for biofuel production is a potential option to generate farm profits and could be an economic incentive for the adoption of cover crops.

14.2.2 Weed Suppression

Herbicide cost is an important component of the overall crop production costs, particularly under no-till management. Use of selective or new herbicides to treat herbicide-resistant weeds has further increased production costs in recent years. In the U.S., the cost of weed control with herbicides is above $11 billion (Kumar et al., 2020). Introduction of cover crops can be part of integrated weed management strategies to manage weeds including herbicide-resistant weeds and thus to reduce costs of herbicide purchase and application (Kumar et al., 2020; Osipitan et al., 2018). Cover crops can reduce weed biomass by 90–100% (Blanco-Canqui et al., 2022).

Grass cover crops such as winter rye and triticale due to lower seed cost and greater biomass production can be more cost-effective for controlling weeds than legume cover crops. Petrosino et al. (2015) reported that one less herbicide application was required to control kochia (herbicide-resistant weed) in winter wheat-fallow systems when cover crops replaced fallow than in those without cover crops. Because herbicides can also be needed to terminate cover crops and thus increase cover crop production costs, termination of cover crops via mowing or harvesting could be an alternative to reduce herbicide application costs provided that cover crops do not regrow. This termination method could deliver a dual service: Reduced herbicide application cost and harvested biomass for livestock or biofuel production.

14.2.3 Nitrogen Credit

Legume cover crops can reduce production costs by reducing the amount of inorganic fertilizer needed for the following crop (Ketterings et al., 2015). For example, a review by Fageria et al. (2005) concluded legume cover crops can fix between 24 and 280 kg ha^{-1} of N$_2$. The benefits of legume cover crops for reducing mineral fertilizer use depends mainly on genetic characteristics and biomass production. Legume cover crops can rapidly release N for the subsequent crop under optimum management and weather conditions for residue decomposition although the recovery of legume-derived N can be lower (<30%) than that from inorganic fertilizer (30–50%; Sievers & Cook, 2018). The monetary value of the amount of N released from cover crops can be equivalent to that of inorganic N fertilizer. The N fertilizer equivalence of legume cover crops for the following crops can range between 12 and 182 kg available N ha^{-1}, with an average of 90 kg N ha^{-1} (Fageria et al., 2005).

The benefits of legume cover crops for increasing crop yields and farm profits depend on the amount of synthetic fertilizer applied. Cover crops can increase crop yields primarily when inorganic fertilization is zero or below recommended fertilization rates. Systems with high price premiums such as organic farming can particularly benefit from cover crops as the primary N source. A case study in a cool temperate region found that corn yields between plots with synthetic fertilization and those with cover crops (crimson clover, hairy vetch, and red clover) did not differ during the three-years transition period to organic farming under corn–soybean–winter wheat rotation (Yang et al., 2019).

Cover crops could reduce the need to purchase synthetic fertilizers, particularly when the recommended N fertilizer rate is low. However, N credits from legume cover crops is not often accounted in systems receiving high amounts of synthetic fertilizers. Proper accounting for the exact amount of N credits from cover crops is required to reduce the total N requirements. In a tropical region, combining synthetic N fertilizer with hairy vetch cover crops by proper adjustment of credits for N from the legume cover crop was a strategy to reduce costs of inorganic fertilization while increasing crop yields (Pes et al., 2022). Thus, combining N from legumes and N from mineral fertilizers could be more profitable than without such combination.

Non-legume cover crops can also accumulate N and reduce production costs by scavenging and reducing losses of N from croplands. For instance, cereal rye can accumulate approximately 100 kg N ha^{-1}, depending on the amount of biomass produced (Fageria et al., 2005). Grass cover crops hold residual N and can slowly release it during the main crop growing season via decomposition. Both legume and non-legume cover crops affect N management in the soil. Legume cover crops decompose more rapidly than non-legumes after termination, and can thus reduce the amount of N fertilizer needed by the following crop, thereby readily reducing production costs.

14.2.4 Soil Carbon Credit

Cover crops can also contribute to farm economics by sequestering C in the soil. On a global scale, cover crops can sequester between 0.21 and 0.56 Mg ha^{-1} yr^{-1}, depending on the information source (Figure 8.1). This amount of C sequestered can be tradable in the developing agricultural C credit market. Increasing the amount of land area under cover crops as well as cover crop biomass production per unit of land can be strategies to further enhance the amount of C sequestered by cover crops and increase thus economic returns. However, it is important to determine the soil C stability under cover crops. While cover crops can increase soil C stocks to qualify for C credits, such increase could rapidly disappear if cover crops are not planted each year after harvest, particularly in degraded or water-limited regions (Blanco-Canqui et al., 2013). Estimating economic returns from C sequestration with cover crops for a given farm based on global averages of C sequestration without accounting the site-specificity of C sequestration can be, however, questionable.

Furthermore, an accurate quantification of cover crop impacts on soil C sequestration in deeper soil depths can be essential for the development of C credits and economic calculations. Most data showing potential of cover crops to accumulate

C are from shallow soil layers (<20 cm depth; Blanco-Canqui, 2022). How cover crops affect deep C storage in the soil profile is still unclear (Tautges et al., 2019). In sum, cover crops, as a potential sink of C, could be used as a valuable C offset tool in current agricultural systems in the emerging C trading programs to generate net returns (Settre et al., 2019).

14.2.5 Crop Residue Harvesting

Another indirect opportunity to increase farm profits from cover crops can be by harvesting residues from the main crops for expanded uses such as livestock or biofuel production and then planting cover crops to offset any negative crop residual removal-induced effects on soil erosion, organic matter loss, and other soil services. In the U.S., corn stover is valued at approximately $60 per ton (Edwards, 2020). Corn stover is harvested and baled from approximately 10% of the total area under corn from corn-producing states in the U.S., but amount of harvested corn stover varies with location (Edwards, 2020). Corn stover harvested can increase as demands for livestock feed and other uses (e.g., biofuel production) increase.

Addition of cover crops could allow greater amounts of corn stover removal for livestock or biofuel production than from corn fields without cover crops. In the U.S. Corn Belt, Pratt et al. (2014) estimated that adding cover crops to corn fields can increase corn stover removal by 4 Mg ha^{-1} relative to fields without cover crops and that combination (corn stover removal followed by cover crops) can be more profitable than cover crops alone, depending on the price of stover. Cover crops can provide much needed cover to maintain soil ecosystem services after crop residue removal while improving farm economics. Cover crops are particularly needed in spring when soils often remain bare before crop planting. While current balers only remove approximately 60% of corn stover in fall after corn harvest, percent surface residue cover the following spring in corn stover-harvested soils can be practically zero due to stover decomposition and losses of stover to wind erosion during winter, which warrants the use of cover crops as a companion practice.

14.2.6 Valuation of Other Ecosystem Services

Cover crops provide many soil ecosystem services that are not often assigned a specific monetary value (Figure 14.1). Such services include reduced soil erosion or sedimentation and improved water quality, biodiversity, and wildlife habitat (e.g., birds, pollinators), among many others. Economic valuation of these benefits needs consideration to enhance farm profitability after cover crop adoption. Placing a monetary value on all soil and environmental services is difficult but can incentivize farmers for greater adoption of cover crops at global scales.

One general question can be: What is the economic value to the whole society for avoided land degradation and improved water and air quality with cover crops? For example, water and wind erosion has on-site and off-site economic costs. On-site economic costs include loss of water, soil, nutrients, and soil C, and a potential reduction in crop yields. Off-site costs include water pollution, sedimentation, and deterioration of air quality, which affect the whole society. A remote sensing approach in the U.S. Corn Belt found that 35% ± 11% of croplands in the region have lost their A horizon, reducing crop yields by 6% ± 2% and causing an economic loss of $2.8 ± $0.9 billion (Thaler et al., 2021). A related example is the loss of costly N from croplands. Estimates across 10 U.S. Midwestern states reported that N loss equals to 1.12 Tg N yr^{-1}, representing an economic loss of $485 million of N fertilizer value (Basso et al., 2019).

Adding cover crops to soils prone to erosion and nutrient leaching can be an effective practice to reduce economic losses. Research data indicate that cover crops can reduce runoff volume by 10–98% and sediment loss by 22–100% (Blanco-Canqui, 2018). Runoff is not clean water but it carries soil, organic matter, and essential nutrients. Thus, growing crops in eroded soils can require purchase of greater amounts of fertilizers than in non-eroded soils to compensate for the loss of nutrient-enriched topsoil. Thus, soil erosion does not only have agronomic and environmental costs but also economic and social costs. If all soil ecosystem services from cover crops are assigned a monetary value, then cover crops can generate economics benefits (Schütte et al., 2020).

14.3 Site-Specificity of Economic Benefits

It is important to discuss that while potential for generating economic benefits from cover crops exists, cover crops may not always improve farm profits (Plastina et al., 2020). This is true especially when cover crops biomass production is low. High amounts of biomass production is needed for erosion control, weed suppression, grazing or harvest, C sequestration, N provision (e.g., legumes), and other services that can improve farm income. The average economic gains with grazing, harvesting, weed suppression, and others discussed above may not be realized in all conditions. It is important to stress that cover crops are a long-term investment.

Economic benefits, especially those from soil-related services (e.g., increased C sequestration), may not surface in the short term (<5 years). Forage production (grazing and harvesting) and weed suppression could offset the costs in the first few years but potential net return can occur in the medium and long term. The need to assess the economic benefits of cover crops on a field or local basis cannot be overemphasized for an accurate economic assessment. In some cases, cost-share or subsidies from state or federal agencies may be necessary to offset costs of cover crop establishment and management and obtain societal benefits from cover crops. Cates et al. (2018) discussed that the probability of negative returns from using cover crops can be between 0% and 77%, depending on cover crop species, mix type, and cover crop biomass production. Also, cost-sharing conditions may not always allow cover crop grazing or harvesting, which need to be considered during cover crop management decision making processes although emerging research data show that cover crop grazing or harvesting may not undo all the intended benefits from cover crops.

14.4 Summary

While a comprehensive economic analysis of cover crops accounting for all soil ecosystem services from cover crops is lacking, the available studies indicate that cover crops have potential to provide positive net returns if biomass production is sufficient. Among the opportunities to generate positive economic outcomes from cover crops include grazing and harvesting cover crops, savings on herbicides and fertilizers, C credits, and monetization of soil ecosystem services. Grazing and harvesting of cover crops can provide immediate positive net returns when biomass production is sufficient under proper grazing (e.g., stocking rates) and harvesting (e.g., cutting height) protocols.

Cover crops can also contribute to farm economics by reducing the amount of fertilizers (e.g., legumes) and herbicides (e.g., weed suppression) and sequestering C (e.g., C credits). Reduced soil erosion, improved water quality, and increased organic matter, and other services from cover crops have not been monetized. Yet, assigning a monetary value to these essential services and social costs can generate further positive net returns and incentivize cover crops adoption. Because cover crop biomass production will affect net returns, maximizing biomass production is imperative.

It is important to discuss that cover crops may not always generate positive net returns in all cases, which warrants consideration of cover crop economics on a local basis. In some cases, cost-sharing can be needed to counterbalance the establishment costs to protect soil and water resources of common societal value. Overall, cover crops can generate income, which may offset the establishment costs and potentially increase net returns while still meeting the original intended soil and environmental goals.

References

Basso, B., Shuai, G., Zhang, J., & Robertson, G. P. (2019). Yield stability analysis reveals sources of large-scale nitrogen loss from the US Midwest. *Scientific Reports, 9*, 5774.

Bergtold, J. S., Ramsey, S., Maddy, L., & Williams, J. R. (2017). A review of economic considerations for cover crops as a conservation practice. *Renewable Agriculture and Food Systems, 34*, 62–76.

Blanco-Canqui, H. (2018). Cover crops and water quality. *Agron. J., 110*, 1633–1647.

Blanco-Canqui, H. (2022). Cover crops and carbon sequestration: Lessons from US studies. *Soil Science Society of America Journal, 86*, 501–519.

Blanco-Canqui, H., Drewnoski, M., & Rice, D. (2021). Does harvesting cover crops eliminate the benefits of cover crops? Insights after three years. *Soil Sci. Soc. Am. J., 85*, 146–157.

Blanco-Canqui, H., Holman, J. D., Schlegel, A. J., Tatarko, J., & Shaver, T. (2013). Replacing fallow with cover crops in a semiarid soil: Effects on soil properties. *Soil Science Society of America Journal, 77*, 1026–1034.

Blanco-Canqui, H., Ruis, S., Holman, H., Creech, C., & Obour, A. (2022). Can cover crops improve soil ecosystem services in water-limited environments? A review. *Soil Science Society of America Journal, 86*, 1–18.

Blanco-Canqui, H., Sindelar, M., & Wortmann, C. S. (2017). Aerial interseeded cover crop and corn residue harvest: Soil and crop impacts. *Agron. J., 109*, 1344–1351.

Cates, A. M., Sanford, G. R., Good, L. W., & Jackson, R. D. (2018). What do we know about cover crop efficacy in the North Central United States? *Journal of Soil and Water Conservation, 73*, 153A–157A.

Dunn, M., Ulrich-Schad, J. D., Prokopy, L. S., Myers, R. L., Watts, C. R., & Scanlon, K. (2016). Perceptions and use of cover crops among early adopters: Findings from a national survey. *Journal of Soil and Water Conservation, 71*, 29–40.

Edwards, W. (2020). Estimating a value for corn stover. Ag Decision Maker. File A1-70. https://www.extension.iastate.edu/AGDm/crops/pdf/a1-70.pdf

Fageria, N. K., Baligar, V. C., & Bailey, B. A. (2005). Role of cover crops in improving soil and row crop productivity. *Communications in Soil Science and Plant Analysis, 36*, 2733–2757.

Florence, A. M., & McGuire, A. M. (2020). Do diverse cover crop mixtures perform better than monocultures? A systematic review. *Agronomy Journal, 112*, 3513–3534.

Franzluebbers, A. J., & Stuedemann, J. A. (2007). Crop and cattle responses to tillage systems for integrated crop–livestock production in the southern Piedmont USA. *Renewable Agriculture and Food Systems, 22*, 168–180.

Holman, J. D., Arnet, K., Dille, J., Maxwell, S., Obour, A., Roberts, T., Roozeboom, K., & Schlegel, A. (2018). Can cover or forage crops replace fallow in the semiarid central Great Plains? *Crop Science, 58*, 932–944.

Ketterings, Q. M., Swink, S. N., Duiker, S. W., Czymmek, K. J., Beegle, D. B., & Cox, W. J. (2015). Integrating cover crops for nitrogen management in corn systems on Northeastern U.S. dairies. *Agronomy Journal, 107*, 1365–1376.

Kumar, A. V., Obour, A., Jha, P., Manuchehri, M. R., Dille, J. A., Holman, J., & Stahlman, P. W. (2020). Integrating cover crops for weed management in the semi-arid U.S. Great Plains: Opportunities and challenges. *Weed Science, 68*, 311–323.

Osipitan, O. A., Dille, J. A., Assefa, Y., & Knezevic, S. Z. (2018). Cover crop for early season weed suppression in crops: systematic review and meta-analysis. *Agron. J., 110*, 2211–2221.

Pes, L. Z., Amado, T. J. C., Gebert, F. H., Schwalbert, R. A., & Pott, L. P. (2022). Hairy vetch role to mitigate crop yield gap in different yield environments at field level. *Scientia Agricola, 79*, e20200327.

Petrosino, J. S., Dille, J. A., Holman, J. D., & Roozeboom, K. L. (2015). Kochia suppression with cover crops in Southwestern Kansas. *Crop For. Turfgrass Manage., 1*, 1–8.

Plastina, A., Liu, F., Miguez, F., & Carlson, S. (2020). Cover crops use in midwestern US agriculture: Perceived benefits and net returns. *Renewable Agriculture and Food Systems, 35*, 38–48.

Pratt, M. R., Tyner, W. E., Jr. Muth, D. J., & Kladivko, E. J. (2014). Synergies between cover crops and corn stover removal. *Agricultural Systems, 130*, 67–76.

Roth, R. T., Ruffatti, M. D., O'Rourke, P. D., & Armstrong, S. D. (2018). A cost analysis approach to valuing cover crop environmental and nitrogen cycling benefits: A Central Illinois on farm case study. *Agricultural Systems, 159*, 69–77.

Schomberg, H., Fisher, D., Reeves, D., Endale, D., Raper, R., Jayaratne, K., Gamble, G., & Jenkins, M. (2014). Grazing winter rye cover crop in a cotton no-till system: Yield and economics. *Agronomy Journal, 106*, 1041–1050.

Schütte, R., Plaas, E., Gómez, J. A., & Guzmán, G. (2020). Profitability of erosion control with cover crops in European vineyards under consideration of environmental costs. *Environment and Development, 35*, 100521.

Settre, C. M., Jeffery, D. C., & Wheeler, S. A. (2019). Emerging water and carbon market opportunities for environmental water and climate regulation ecosystem service provision. *Journal of Hydrology, 578*, 124077.

Sievers, T., & Cook, R. L. (2018). Aboveground and root decomposition of cereal rye and hairy vetch cover crops. *Soil Science Society of America Journal, 82*, 147–155.

Siri-Prieto, G., Wayne Reeves, D., & Raper, R. L. (2007). Tillage systems for a cotton–peanut rotation with winter-annual grazing: Impacts on soil carbon, nitrogen and physical properties. *Soil and Tillage Research, 96*, 260–268.

Tautges, N. E., Chiartas, J. L., Gaudin, A. C. M., O'geen, A. T., Herrera, I., & Scow, K. M. (2019). Deep soil inventories reveal that impacts of cover crops and compost on soil carbon sequestration differ in surface and subsurface soils. *Global Change Biology, 25*, 3753–3766.

Thaler, E. A., Larsen, I. J., & Yiu, Q. (2021). The extent of soil loss across the US Corn Belt. *Proceedings of the National Academy of Sciences of the United States of America, 118*, e1922375118.

Tobin, C., Kumar, S., Wang, T., & Sexton, P. (2020). Demonstrating short-term impacts of grazing and cover crops on soil health and economic benefits in an integrated crop-livestock system in South Dakota. *Open Journal of Soil Science, 10*, 109–136.

Yang, X., Drury, C., Reynolds, W., & Reeb, M. (2019). Legume cover crops provide nitrogen to corn during a three-year transition to organic cropping. *Agronomy Journal, 111*, 3253–3264.

15

Adaptation to Extreme Weather

15.1 Extreme Weather Events

The frequent droughts, floods, heat waves, and erratic rainfalls and snowfalls we have been experiencing are examples of extreme weather events on a global scale (Robinson, 2021). Severe droughts or heat waves are often followed by intense localized rainstorms and floods in subsequent years, reflecting the abnormal and fluctuating conditions. These extreme weather events are unfortunately becoming a norm rather than isolated events (Kaye & Quemada, 2017). Some of these extremes have delayed timely field operations (e.g., crop planting), altered growing seasons, impeded crop germination, and reduced crop yields and farm economics in recent decades.

Climate modeling exercises indicate further weather fluctuations in the future with more erratic rainfalls, higher temperatures, recurrent droughts, and other extreme scenarios (Robinson, 2021). However, one does not have to travel far or only follow future projections from computer models to believe in extreme events. The past and current extreme weather events we have seen are enough reasons to design new or redesign current soil management strategies to adapt to such events. The key question is: How can we build the resilience of soils to adapt to such extreme events? Can cover crops be a strategy to enhance soil resilience to adapt to extreme weather? Adapting to fluctuating weather conditions means adapting to drier, wetter, cooler, and warmer conditions.

How cover crops can contribute to adaptation to extreme weather events has not been discussed at length. Kaye and Quemada (2017), using two case studies in central Spain and Pennsylvania, U.S., discussed that cover crops can contribute to climate change mitigation and adaptation while still providing traditional ecosystem services. This chapter considers how cover crops can contribute to the

Cover Crops and Soil Ecosystem Services, First Edition. Humberto Blanco.
© 2023 American Society of Agronomy, Inc. / Crop Science Society of America, Inc.
/ Soil Science Society of America, Inc. Published 2023 by John Wiley & Sons, Inc.

management of fluctuations in weather such as droughts, floods, extreme precipitation events, dust storms, and extreme soil temperatures.

15.2 Droughts

Specific research on how cover crops affect agroecosystem drought resistance is still limited. The emerging studies on the topic discuss that diversifying crop rotations with cover crops can be a strategy to enhance the resilience of agricultural lands against droughts and other extreme events (Renwick et al., 2021). The decrease in crop diversity and the increase in homogenous or simplified rotations have reduced overall drought resistance of agricultural systems. Adding cover crops can diversify current simplified systems and reduce their sensitivity to water stress and increase their overall resilience. Different plant species can differently respond and adapt to fluctuating environmental and climatic conditions. A drought experiment using rainout shelters imposed on 36-years of corn-based rotations with and without a red clover cover crop in a temperate environment found that addition of a cover crop enhanced corn drought resistance, maintained yields, and reduced drought-induced losses in crop yield by 17.1% (Renwick et al., 2021).

Cover crops can have positive and negative effects on soil water. Cover crops use water and can deplete water for the next crop in both water-limited regions and during dry years in non-water limited regions. However, while growing cover crops reduce water for the subsequent crops, cover crop residues can contribute to water storage after termination by reducing soil temperature and evaporation. A review focusing on water-limited regions found that cover crops reduce soil water in approximately 50% of cases (Blanco-Canqui et al., 2022). This suggests that cover crops under appropriate management may not always result in water depletion in water-limited regions. Cover crops may not reduce soil water storage compared with no cover crops, depending on the climatic region. During the 2012 drought in North America, winter rye cover crop did not significantly reduce soil-profile water storage relative to no cover crop at three sites under corn–soybean rotation in Iowa and Indiana, U.S., with a mean annual precipitation >800 mm (Daigh et al., 2014).

Cover crops could also contribute to drought resistance by improving soil properties such as soil aggregate stability, macroporosity, organic matter concentration, and water infiltration in the long term, although cover crop effects on plant available water content are mixed. A global review found that cover crops may or may not increase plant available water content (Blanco-Canqui & Ruis, 2020). Even in the long term (>10 years), cover crop effects on plant available water content can be inconsistent. Also, while it is often thought that an increase in soil organic matter concentration with cover crops can increase soil water holding capacity, experimental data show that even significant increases in soil organic matter concentration can only have minimal effects on plant available water (Minasny & McBratney, 2018). However, while cover crop effects on plant available water may be mixed, the improved soil structural properties with cover crops could allow proliferation of roots to deeper depths to acquire water (Hunter et al., 2021).

Research suggests cover crops can primarily contribute to drought resistance by reducing evaporation via residues left on the soil surface under no-till systems. The extent to which cover crops can reduce drought stress depends on the severity of drought and climatic region. Cover crop residue mulch may be more effective at conserving soil water under relatively mild or moderate droughts than under severe droughts. Further, in water-limited environments, cover crops can amplify drought impacts by depleting the limited amount of water. Planting cover crops only when precipitation is at or above normal can be a strategy to avoid aggravating drought impacts on crop production. Developing climate models that can accurately forecast precipitation input in advance of the next crop season can be valuable to manage cover crops (Robinson, 2021).

Cover crop management such as termination timing, time after introduction, and tillage system could affect how cover crops impact drought resistance. Timely termination of cover crops can be critical to allow water recharge and offset water depletion before planting crops in water-limited regions (Baxter et al., 2021). Also, cover crop residue mulch combined with no-till management can protect soil and conserve water relative to tilled systems where cover crop residues are incorporated into the soil (Hunter et al., 2021). Hunter et al. (2021) reported that single species cover crops (red clover, winter rye, and forage radish) and a mix (red clover, winter rye, and Austrian pea) did not reduce corn drought stress in corn silage–soybean–winter wheat rotation when cover crops were incorporated into the soil approximately two to three weeks before planting corn. Overall, research data show that cover crops have the potential to improve resilience of agricultural lands against droughts by conserving soil and water but such potential will depend on climatic region, drought severity, and cover crop management.

15.3 Floods

Similar to drought, flooding is another extreme event that has major adverse agronomic and environmental implications. Flooding risks have increased in recent decades due to localized and intense precipitation events (Kaur et al., 2020). Occasional floods and waterlogged conditions are not uncommon in agricultural lands with nearly level and fine-textured soils, particularly in springtime (Figure 15.1). Aeration, nutrient uptake, and root respiration and growth, and other processes are rapidly compromised when all the soil pores are filled with water (soil matric potential = 0) and all or a portion of the plant is submerged in water. While intense precipitation is the main cause of flooding, management-induced causes such as increased soil compaction and reduced soil macroporosity, drainage, and infiltration can further accentuate risks for flooding.

Can cover crops reduce risks of flooding? Cover crops can be a potential strategy to reduce flooding by (a) using excess soil water and (b) improving soil properties although cover crop establishment can be difficult when the soil is too wet. Cover crops can consume significant amounts of extra soil water during the growing season and potentially dry out saturated fields between harvest and planting. Excessive soil water consumption by cover crops can be a problem in water-limited regions, but it can be highly beneficial in flood-prone environments. In a temperate region with high rainfall and snowmelt input, Kahimba et al. (2008) found a berseem clover cover crop altered the soil-profile distribution of soil water and removed excess water from the crop root zone. Deep-rooted and high-biomass producing cover crops can be an alternative or a companion practice to surface and tile drainage systems to remove excess water and reduce waterlogging.

Figure 15.1 Flooded cropland under intense rainstorms in the spring near Clay Center, NE, U.S. Photo by H. Blanco.

Mechanical drainage systems can remove excess water from the root zone but can also serve as conduits for transport of nutrients and other chemicals to groundwater. In turn, cover crops can reduce excess water and reduce flooding risks while adsorbing nutrients and reducing non-point source pollution.

Cover crops can also improve drainage by increasing water infiltration and improving other soil hydraulic properties. A review concluded that cover crops can increase water infiltration by an average of 43% (Blanco-Canqui & Ruis, 2020). Cover crops alter the soil-pore size distribution, increasing the proportion of macropores, and promoting water flow to deeper layers. While changes in soil pore-size distribution and other soil properties can manifest in the long term, well-established deep-rooted cover crops can open up compacted layers and increase water infiltration to deeper layers in the profile in the short term (Chen & Weil, 2011).

It is important to note that while cover crops can contribute to the management of waterlogged soils or moderately flooded soils, their effectiveness can be limited under severe and extended floods. Effectiveness of cover crops for reducing flood risk, similar to other soil ecosystem services, can depend on cover crop management. It can increase with the increase in the amount of biomass produced, years after adoption, and rooting depth. Research data on the magnitude to which cover crops can reduce flooding risks or alleviate waterlogging under fluctuating climates are needed to further evaluate the extent of cover crop benefits for managing floods.

15.4 Precipitation Extremes

Extreme precipitation events are also a major concern similar to droughts and floods (Kahraman et al., 2021; Kaye & Quemada, 2017). Rainfall intensities can be classified in low or moderate ($<20\,\mathrm{mm\,h^{-1}}$) and high ($20–50\,\mathrm{mm\,h^{-1}}$) (Sastre et al., 2016). However, rainfall intensities above $50\,\mathrm{mm\,h^{-1}}$ are not uncommon under current climatic scenarios. Such intense rainstorms can cause severe soil erosion and flooding, especially in spring when soils have limited protective cover (Figure 15.1). Intense rainstorms can be frequent even when the total annual precipitation amount among years remains unchanged (Kahraman et al., 2021). Furthermore, models estimate that water erosion may increase by as much as 22% in Europe by 2050 due to the projected increases in rainfall erosivity (Panagos et al., 2021).

Severe soil erosion is a response to precipitation extremes and depends more on rainfall intensity than on rainfall amount. Losses of soil, water, and nutrients accelerate as the rainfall erosivity increases. Because of its greater kinetic energy than low or moderate intensity rainfall, an intense rainstorm rapidly seals and consolidates the surface, disintegrates soil aggregates, splashes soil particles, and accelerates transport of nutrient-enriched soil. The higher the kinetic energy of rainfall, the greater the amount of detached soil particles available to runoff. The occurrence of frequent intense rainstorms may not be manageable, but what is manageable is the soil protective cover against the erosive force of intense storms.

Can cover crops buffer impacts from precipitation extremes? Cover crops can be a strategy to adapt to extreme rainfall events and reduce erosion. Available data under simulated rainfall at high intensities show cover crops can reduce runoff by 10–89% and soil loss by 39–96% under 63.5–125 mm h^{-1} of rainfall (Table 15.1). These result show cover crops can be effective at reducing runoff and sediment loss even under very high intensity rainfalls. The simulated rainfall intensities (63.5–125 mm h^{-1}) are approximately twice the typical high rainfall intensities (20–50 mm h^{-1}) under natural conditions. Table 15.1 also indicates that cover crop effectiveness for managing erosion under intense rainstorms can be highly variable, depending on cover crop biomass production and cover crops species.

As with other soil ecosystem services, the positive effects of cover crops on adapting to intense rainstorms can be minimal if aboveground and belowground

Table 15.1 Cover Crops can be Effective at Reducing Runoff and Losses of Sediment at High Rainfall Intensities

Location	Soil slope (%)	Cover crop	Rain intensity (mm h^{-1})	Runoff reduction (%)	Sediment reduction (%)
[a]New York, U.S.	5	Winter rye	na	46	73
[b]Iowa, U.S.	0.1–2.7	Winter rye	65	65	68
[c]Wisconsin, U.S.	8–15	Winter rye	70	54	55
[d]Kansas, U.S.	1–3	Triticale	63	70	70
[e]Wisconsin, U.S.	2.2	Red clover	70	43	96
		Italian ryegrass		69	83
		Winter rye		89	96
[f]Iowa, U.S.	2.8–6	Rye	125	10	86
		Oat		0	42

[a] Griffith et al. (2020).
[b] Korucu et al. (2018).
[c] Siller et al. (2016).
[d] Blanco-Canqui et al. (2013).
[e] Grabber and Jokela (2013).
[f] Kaspar et al. (2001).
na = not available.

cover crop biomass production is limited. For example, in temperate regions, cover crops planted after corn or soybean harvest in late fall do not establish well due to low temperatures and provide limited cover during periods of high rainfall intensity (late winter or early spring). Establishing high-biomass producing cover crops such as winter rye in temperate regions ahead of intense rainfall seasons can be a strategy to manage soil response to extreme rainstorms.

Selection of cover crop species can be important for mitigating soil erosion and flood risks under intense precipitation events. Cover crop species that overwinter (e.g., winter hardy grasses) can reduce runoff and soil loss in spring more than non-winter hardy species. In the U.S. Midwest, under high-intensity simulated rainfall, Kaspar et al. (2001) observed that a winter rye cover crop reduced runoff volume and rill erosion more than a oat cover crop crop due to better survival in winter and greater biomass production in spring. Similarly, in the central U.S. Great Plains, under intense simulated rainfall, triticale cover crop reduced runoff and sediment loss more than legume cover crops due to its higher biomass production (Blanco-Canqui et al., 2013). Winter rye, ryegrass, triticale, wheat, and barley are some of the overwintering grass species that can produce significant amounts of biomass under proper management.

Cover crops can protect soil under intense rainfalls by at least three mechanisms. One, cover crops (canopy and residues) can intercept raindrops and buffer their erosive energy. Soil erosion increases as the kinetic energy of rainstorms increases and exponentially decreases as the amount of surface crop residue increases. Greater than 70% cover crop canopy or residue cover can be needed to reduce erosion under intense rainstorms. Two, soils under cover crops are more resistant to rainfall erosive energy due to the higher soil aggregate stability and organic matter concentration compared with soils without cover crops (Blanco-Canqui & Ruis, 2020). Three, cover crop roots stabilize soil and such stabilization can be particularly beneficial for reducing concentrated runoff, which causes rill and ephemeral gully erosion. Soil resistance to erosion decreases exponentially as the amount of cover crop root biomass increases (Gyssels et al., 2005).

Available data suggest soils with cover crops can be less susceptible and more resilient to erosion than bare soils. Well-established cover crops can provide immediate protective cover to soil from erosion although soil properties that infer resistance to erosion including soil aggregate stability, organic matter concentration, and water infiltration can be slow to develop. Also, cover crops not only reduce water erosion but also reduce nutrient leaching under intense rainfalls (Korucu et al., 2018). Cover crops take up nutrients such as nitrates and reduce their availability for runoff. Overall, adding cover crops can be an important tool to help agricultural soils to adapt to precipitation extremes.

15.5 Dust Storms

Frequent dust storms are also a concern under extreme weather conditions, particularly in arid and semiarid regions. Dust storms can be reduced to negligible levels if soil is covered year round with growing vegetation or crop residues. The limited residue cover combined with droughts and lack of adaptation to extreme events led to the Dust Bowl in the U.S. in the 1930s (Bennett, 1939). Presently, droughts are not uncommon due to variable or abnormal precipitation. Dust storm events could increase further under fluctuating weather patterns with recurrent droughts.

Available data indicate that cover crops combined with no-till practices can be one of the tools to reduce wind erosion. Under wind speeds as high as $50\,km\,h^{-1}$, in the arid southwest U.S., Japanese millet, pearl millet, brown top millet, and sorghum-sudangrass cover crops reduced soil loss by wind erosion more than no cover crop and their effectiveness rapidly increased as days after establishment increased (Darapuneni et al., 2021). The same study found sorghum-sudangrass cover crop reduced wind erosion more than other cover crops due to its greater surface coverage and biomass production.

High-biomass producing cover crops can be an effective tool to adapt to wind storms. Cover crops anchor the soil, shield the soil surface, and reduce dust generation (Van Pelt & Zobeck, 2004). Also, cover crops reduce the soil's susceptibility to wind erosion by improving soil dry aggregate properties in wind-erosion affected regions (Blanco-Canqui et al., 2013). Establishing cover crops in wind-erosion prone environments can be, however, a challenge due to limited precipitation. Using limited irrigation or one or two irrigation events during establishment can help with implementing cover crops when precipitation is limited during establishment. Cover crops can complement windbreaks and no-till systems to adapt to high winds and reduce dust storms.

15.6 Temperature Extremes

Extreme high or low temperatures are also indicators of fluctuating weather patterns. Extreme heat and cold events are not uncommon under current weather scenarios and their attendant adverse impacts on crop production. For example, heat waves, which refer to short periods of high temperature, can cause heat stress in crops, increase the demand for water, damage plant metabolic processes, accelerate crop phenological stages, and reduce crop yields (Vogel et al., 2019). Also, extreme low temperatures can reduce crop establishment, organic matter decomposition, biological activity, seed emergence, and alter other processes.

Can cover crops buffer the extreme high or low temperatures in the soil? A review found that cover crops can change daytime and nighttime temperature as well as seasonal soil temperatures (Blanco-Canqui et al., 2020). On average, soils under cover crops can be 2 °C cooler during the day and 1 °C warmer at night relative to no cover crops. On a seasonal basis, soils under cover crops are warmer in winter and cooler during the rest of the year. Cover crops buffer, moderate, and stabilize soil temperature. Cover crops can also alter the amplitude of soil temperature by reducing the maximum and increasing minimum soil temperatures. The significant reduction in maximum and minimum soil temperatures indicates that cover crops could moderate extreme high or low temperatures and thus reduce the potential adverse impacts of abrupt fluctuations in soil temperature.

Cover crops can alter solar radiation at the soil surface and thus alter soil temperature via their canopy cover or residues. Cover crops are like a blanket to soil and can create cooler or warmer microclimates, depending on the season. Crop residues often cover the soil only during a portion of the year. Addition of cover crops can augment surface cover to moderate soil temperature until main crops are established. Presence of abundant cover crop residues in spring could delay soil warming and seed germination, but such delay is compensated by increased crop growth in summer due to reduced evaporation and abundant soil water.

Cover crops can also alter soil thermal properties related to soil microclimate including volumetric heat capacity and thermal diffusivity although studies on these soil properties are few (Haruna et al., 2017). Volumetric heat capacity refers to the change in heat content of the soil per unit change in temperature, while thermal diffusivity refers to the rate of heat transmission through the soil. Cover crops can increase volumetric heat capacity and reduce thermal diffusivity, which indicates cover crops can reduce abrupt or rapid changes in heat flow within the soil compared with no cover crops. Cover crops affect soil thermal properties mostly through their effects on soil water (Sindelar et al., 2019). The relatively stable soil temperature under cover crops can be important for buffering extreme high and low air temperatures at the soil-atmosphere interface.

15.7 Soil Resilience

Soils are under increasing pressures from droughts, floods, heat waves, and other extreme weather events. Such pressure is often accentuated by frequent soil disturbance, reduced residue cover, lack of crop diversity, lack of continuous living plants, and contamination (e.g., agrichemicals), among others. The ability of a soil to recover from such external and internal stresses and still provide the intended ecosystem services is considered here as resilience (Hunter et al., 2021;

Ludwig et al., 2018). Resilient soils should be able to rebound, possess a spring-like behavior, and rapidly return to pre-perturbation levels. Changes in soil physical, chemical, and biological processes and properties are indicators of how resistant a soil is to increasing perturbations and climatic fluctuations. Soil resilience is not a simple concept as it relates to a holistic view of soil and its adaptive capacity to stresses for continued delivery of essential ecosystem services (Ludwig et al., 2018).

Cover crops can be a strategy for agricultural lands to develop resilience and adapt to anthropogenic perturbations and fluctuating climates. As discussed earlier, cover crops can reduce soil erosion under intense rains or windstorms, reduce excessive evaporation during droughts, reduce flood risks, and moderate extreme soil temperature, thereby contributing to the overall resilience of agricultural soils (Table 15.2). In the long term, cover crops can improve the rebounding capacity of the soil by enhancing the physical, chemical, and biological properties of the soil. Addition of cover crops to perturbed systems has been found to improve soil structural and hydraulic properties, restore some of the C lost, activate biological activities, and improve other parameters of soil resilience (McClelland et al., 2021; Kim et al., 2020).

Table 15.2 Mechanisms by which Cover Crops can Enable Agricultural Soils to Develop Resilience and Adapt to Increasing Extreme Weather Events

Extreme event	Cover crop benefits
Drought	• Reduce evaporation and conserve water • Increase macroporosity and water infiltration • Increase soil organic matter concentration • Increase water holding capacity in some cases • Improve rooting depth to access water in deeper profile
Flood	• Reduce excess water by transpiration • Improve soil macroporosity • Improve water infiltration and hydraulic conductivity • Improve soil rebounding capacity after floods
Intense rainstorms	• Protect the soil surface from raindrop impacts • Promote water infiltration and delay runoff initiation • Reduce soil surface crusting and runoff generation • Improve soil structural properties (e.g., improved wet aggregate stability) • Increase soil organic matter concentration in the long term • Anchor the soil with roots and reduce formation of concentrated runoff

(Continued)

Table 15.2 (Continued)

Extreme event	Cover crop benefits
Dust storms	• Increase soil surface roughness • Intercept and buffer wind energy • Trap and reduce blowing dust • Reduce soil erodibility (e.g., improved dry aggregate stability) • Anchor the soil with roots and reduce soil particle movement
Extreme temperatures	• Cover and insulate the soil surface • Reduce evaporation and conserve water, which alter thermal properties • Increase soil organic matter concentration • Stabilize and moderate transport of heat within the soil

One of the primary attributes by which cover crops can contribute to soil resilience is by increasing soil organic matter levels although cover crops may not always increase organic matter concentration, particularly in the short term. One of the examples of how cover crop-induced increase in soil organic matter enhances soil resilience is by improving soil's resistance to erosion and compaction (Blanco-Canqui et al., 2011). Because organic matter promotes soil aggregation and imparts elastic behavior to soil, soils under cover crops can be less erodible and rebound more rapidly after compaction. Soils under cover crops could recover from stresses more rapidly than those without cover crops. Soils covered with cover crops can be less vulnerable and more resistant to extreme weather events than those without cover crops (Kaye & Quemada, 2017). Overall, cover crops should be put at the top in the agenda for enhancing resilience of agricultural lands under fluctuating weather patterns.

15.8 Summary

Cover crops have potential to contribute to adaptation to droughts, floods, heat waves, intense rainstorms, dust storms, and other extreme events under fluctuating weather conditions while delivering the traditional ecosystem services. Cover crops can contribute to the adaptation of cropping systems to fluctuating weather patterns. While cover crops use water and may reduce crop yields in water-limited regions, cover crop residues can improve drought resistance, to some degree, by reducing evaporation, conserving water, and improving soil properties.

Cover crops can also be a companion practice to drainage systems to manage floods or waterlogging by using excess water through transpiration, and by improving water infiltration. Reducing both water erosion under high-intensity rainfalls and wind erosion under high wind speeds is another benefit from cover crops. Well-established cover crops can complement windbreaks to manage wind erosion. Cover crops can moderate abrupt fluctuations in temperature and reduce the temperature amplitude by reducing the maximum and increasing minimum soil temperatures, and developing microclimates.

Cover crop effectiveness depends, however, on cover crop biomass production, cover crop species, tillage systems, and the extent of weather fluctuations. High-biomass producing cover crops coupled with no-till management can be an adaptation strategy to fluctuating climatic conditions. One of the mechanisms by which cover crops could improve soil resilience against extreme events can be by increasing soil organic matter concentration. More specific research data on the extent to which cover crops can help with adaptation to extreme climatic events are needed. Available information indicates cover crops can be one of the tools within the portfolio of options to develop agroecosystem resilience against climatic fluctuations.

References

Baxter, L. L., West, C. P., Brown, C. P., & Green, P. E. (2021). Cover crop management on the southern high plains: Impacts on crop productivity and soil water depletion. *Animal (Basel), 11*, 212.

Bennett, H. H. (1939). *Soil conservation* (993 p.). McGraw-Hill Book Co., Inc.

Blanco-Canqui, H., Holman, J. D., Schlegel, A. J., Tatarko, J., & Shaver, T. (2013). Replacing fallow with cover crops in a semiarid soil: Effects on soil properties. *Soil Science Society of America Journal, 77*, 1026–1034.

Blanco-Canqui, H., Mikha, M., Presley, D. R., & Claassen, M. M. (2011). Addition of cover crops enhances no-till potential for improving soil physical properties. *Soil Science Society of America Journal, 75*, 1471–1482.

Blanco-Canqui, H., & Ruis, S. (2020). Cover crops and soil physical properties. *Soil Science Society of America Journal*, 1527–1576.

Blanco-Canqui, H., Ruis, S., Holman, H., Creech, C., & Obour, A. (2020). Can cover crops improve soil ecosystem services in water-limited environments? A review. *Soil Science Society of America Journal, 86*, 1–18.

Blanco-Canqui, H., Ruis, S. J., Holman, H. D., Creech, C., & Obour, A. (2022). Can cover crops improve soil ecosystem services in water-limited environments? A Review. *Soil Science Society of America Journal, 86*, 1–18.

Chen, G., & Weil, R. R. (2011). Root growth and yield of maize as affected by soil compaction and cover crops. *Soil and Tillage Research, 117*, 17–27.

Daigh, A. L., Helmers, M. J., Kladivko, E., Zhou, X., Goeken, R., Cavdini, J., Barker, D., & Sawyer, J. (2014). Soil water during the drought of 2012 as affected by rye cover crops in fields in Iowa and Indiana. *Journal of Soil and Water Conservation, 69*, 564–573.

Darapuneni, M. K., Idowu, O. J., Sarihan, B., DuBois, D., Grover, K., Sanogo, S., Djaman, K., Lauriault, L., Omer, M., & Dodla, S. (2021). *Applied Engineering in Agriculture, 37*, 11–23.

Grabber, J. H., & Jokela, W. E. (2013). Off-season groundcover and runoff characteristics of perennial clover and annual grass companion crops for no-till corn fertilized with manure. *Journal of Soil and Water Conservation, 68*, 411–418.

Griffith, K. E., Young, E. O., Klaiber, L. B., & Kramer, S. R. (2020). Winter rye cover crop impacts on runoff water quality in a northern New York (USA) tile-drained maize agroecosystem. *Water, Air, & Soil Pollution*, 231–284.

Gyssels, G., Poesen, J., Bochet, E., & Li, Y. (2005). Impact of plant roots on the resistance of soils to erosion by water: A review. *Progress in Physical Geography, 29*, 189–217.

Haruna, S. I., Anderson, S. H., Nkongolo, N. V., Reinbott, T., & Zaibon, S. (2017). Soil thermal properties influenced by perennial biofuel and cover crop management. *Soil Science Society of America Journal, 81*, 1147–1156.

Hunter, M. C., Kemanian, A. R., & Mortensen, D. A. (2021). Cover crop effects on maize drought stress and yield. *Agriculture, Ecosystems and Environment, 311*, 107294.

Kahimba, F. C., Ranjan, R. S., Froese, J., Entz, M., & Nason, R. (2008). Cover crop effects on infiltration, soil temperature, and soil moisture distribution in the Canadian prairies. *Applied Engineering in Agriculture, 24*, 321–333.

Kahraman, A., Kendon, E. J., Chan, S. C., & Fowler, H. J. (2021). Quasi-stationary intense rainstorms spread across Europe under climate change. *Geophysical Research Letters, 48*, e2020GL092361.

Kaspar, T. C., Radke, J. K., & Laflen, J. M. (2001). Small grain cover crops and wheel traffic effects on infiltration, runoff, and erosion. *Journal of Soil and Water Conservation, 56*, 160–164.

Kaur, G., Singh, G., Motavalli, P. P., Nelson, K. A., Orlowski, J. M., & Golden, B. R. (2020). Impacts and management strategies for crop production in waterlogged/flooded soils: A review. *Agronomy Journal, 112*, 1475–1501.

Kaye, J., & Quemada, M. (2017). Using cover crops to mitigate and adapt to climate change. A review. *Agronomy for Sustainable Development, 37*, 4.

Kim, N., Zabaloy, M. C., Guan, K., & Villamil, M. B. (2020). Do cover crops benefit soil microbiome? A meta-analysis of current research. *Soil Biology and Biochemistry, 142*, 107701.

Korucu, T., Shipitalo, M. J., & Kaspar, T. C. (2018). Rye cover crop increases earthworm populations and reduces losses of broadcast, fall-applied, fertilizers in surface runoff. *Soil and Tillage Research, 180*, 99–106.

Ludwig, M., Wilmes, P., & Schrader, S. (2018). Measuring soil sustainability via soil resilience. *Science of the Total Environment, 626*, 1484–1493.

McClelland, S. C., Paustian, K., & Schipanksi, M. E. (2021). Management of cover crops in temperate climates influences soil organic carbon stocks - a meta-analysis. *Ecological Applications, 31*, e02278.

Minasny, B., & McBratney, A. B. (2018). Limited effect of organic matter on soil available water capacity. *European Journal of Soil Science, 69*, 39–47.

Panagos, P., Ballabio, C., Himics, M., Scarpa, S., Matthews, F., Bogonos, M., Poesen, J., & Borrelli, P. (2021). Projections of soil loss by water erosion in Europe by 2050. *Environmental Science & Policy, 124*, 380–392.

Renwick, L. L. R., Deen, W., Silva, L., Gilbert, M. E., Maxwell, T., Bowles, T. M., & Gaudin, A. C. M. (2021). Long-term crop rotation diversification enhances maize drought resistance through soil organic matter. *Environmental Research Letters, 16*, 084067.

Robinson, W. A. (2021). Climate change and extreme weather: A review focusing on the continental United States. *Journal of the Air and Waste Management Association, 71*, 1186–1209.

Sastre, B., Barbero-Sierra, C., Bienes, R., Marques, M. J., & García-Díaz, A. (2016). Soil loss in an olive grove in Central Spain under cover crops and tillage treatments, and farmer perceptions. *Journal of Soils and Sediments, 17*, 873–888.

Siller, A. R. S., Albrecht, K. A., & Jokela, W. E. (2016). Soil erosion and nutrient runoff in corn silage production with Kura clover living mulch and winter rye. *Agronomy Journal, 108*, 989–999.

Sindelar, M., Blanco-Canqui, H., Virginia, J., & Ferguson, R. (2019). Do cover crops and corn residue removal affect soil thermal properties? *Soil Science Society of America Journal, 83*, 448–457.

Van Pelt, R. S., & Zobeck, T. M. (2004). *Effects of polyacrylamide, cover crops, and crop residue management on wind erosion.* ISCO – 13th International Soil Conservation Organization Conference Conserving Soil and Water for Society: Sharing Solutions. Brisbane, Australia.

Vogel, E., Donat, M. G., Alexander, L. V., Meinshausen, M., Ray, D. K., Karoly, D., & Frieler, K. (2019). The effects of climate extremes on global agricultural yields. *Environmental Research Letters, 14*, 054010.

16

Opportunities, Challenges, and Future of Cover Crops

16.1 Opportunities

Cover crops can be multi-functional under proper management based on the available research information discussed in previous chapters. Cover crops can thus offer an opportunity to restore, maintain, and/or improve soil ecosystem services, and potentially increase farm profitability and overall soil adaptation to fluctuating weather patterns (Figure 16.1). They can be an important companion practice to other conservation practices such as no-till, diversified crop rotations, and conservation buffers. Cover crops and other conservation practices can complement each other and enhance their performance through synergistic effects.

16.1.1 Ecosystem Services

Experimental data indicate cover crops can generally improve soil physical, chemical, and biological properties, reduce water and wind erosion, reduce nutrient leaching, increase C sequestration and soil fertility, suppress weeds, and provide opportunities for grazing and harvesting (Table 16.1). Cover crops can be particularly beneficial for reducing soil erosion, retaining nutrients, and suppressing weeds including herbicide-resistance weeds (Adetunji et al., 2020; Osipitan et al., 2018). In the long term, cover crops can also restore some of the lost soil C by removing CO_2 from the atmosphere and storing it in the soil (McClelland et al., 2021; Poeplau & Don, 2015). Research shows cover crops can, on average, sequester significant amounts of C in the soil in the long term when biomass production is consistently high ($>1\,Mg\,ha^{-1}$).

Cover Crops and Soil Ecosystem Services, First Edition. Humberto Blanco.
© 2023 American Society of Agronomy, Inc./Crop Science Society of America, Inc.
/Soil Science Society of America, Inc. Published 2023 by John Wiley & Sons, Inc.

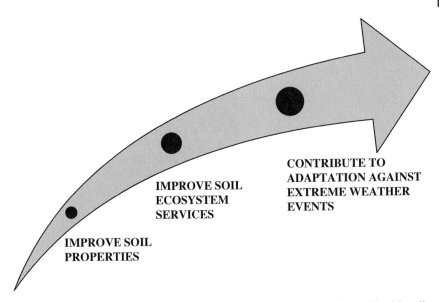

Figure 16.1 Cover crops can be multi-functional systems for managing soil health, soil ecosystem services, and present and future extreme weather scenarios.

Table 16.1 Opportunities and Challenges of Cover Crops for the Management of Soil Ecosystem Services

Ecosystem services	Cover crop opportunities	Cover crop challenges
Biomass production	• $3.78 \pm 3.08\,\mathrm{Mg\,ha^{-1}}$ in temperate regions • $2.40 \pm 2.33\,\mathrm{Mg\,ha^{-1}}$ in semiarid regions • $> 4\,\mathrm{Mg\,ha^{-1}}$ in subtropical and tropical regions	• Highly variable • Lower in colder than in warmer regions
Soil health	• Improve soil properties	• Changes mostly measurable in $>10\,\mathrm{yr}$ • Changes occur mostly in $<20\,\mathrm{cm}$ depth
Water erosion	• Reduce soil loss and runoff • Reduce in this order: Sediment loss > Runoff > Nutrient runoff	• Runoff may not always decrease

(Continued)

Table 16.1 (Continued)

Ecosystem services	Cover crop opportunities	Cover crop challenges
Wind erosion	• Reduce wind erosion	• May not reduce soil erodibility by wind
Nutrient loss	• Reduce nutrient leaching	• Limited effect on dissolved nutrients • Increase dissolved nutrients in runoff
Greenhouse gas emissions	• Non-legumes reduce N_2O emissions	• Increase CO_2 emissions • Limited effect on CH_4 emissions • Legumes increase N_2O emissions
Carbon sequestration	• 0.21–0.56 Mg C ha^{-1} yr^{-1} in <30 cm depth	• Low if biomass is <1 Mg ha^{-1} and <5 yr of adoption • Increase C mostly in the 0–20 cm soil depth
Soil water	• Improve drainage • Reduce evaporation and conserve water	• Deplete water in 50% of cases in water-limited regions • Mixed effects on plant available water
Weed management	• Reduce early crop season weeds • Suppress weeds by 90–100%	• Limited biomass production
Soil fertility	• Legumes provide high amounts of N • Improve soil chemical properties	• Can be a source and sink of P • Limited effect in fertile soils
Crop yields	• Legumes can increase crop yields	• Non-legumes could reduce crop yields • Legumes have less beneficial effects in cool than warmer in temperate regions
Grazing and harvesting	• Grazing and harvesting are viable	• Can compact soil in some cases
Economics	• Can generate positive net returns	• Difficult to value all ecosystem services • May not generate net returns in all cases
Extreme weather	• Contribute to adaptation to extreme weather	• Dependent on cover crop management

Cover crops can also contribute to improved nutrient use efficiency by reducing upward (e.g., N_2O emissions), lateral (e.g., associated with eroded sediment), and downward (e.g., leaching) losses of nutrients. Non-legume cover crops can scavenge nutrients, thereby reducing the nutrient availability for losses (e.g., leaching). Legume cover crops can fix significant amounts of N, which can potentially reduce N fertilizer requirements for main crops. Synchronizing nutrient release from cover crops and nutrient uptake by main crops can be an opportunity to reduce nutrient losses and save on fertilizer purchases. While non-legume cover crops do not often increase crop yields, legume cover crops can increase crop yields by providing available N (Abdalla et al., 2019; Marcillo & Miguez, 2017). Such an increase in crop yields is especially high when limited or no inorganic fertilizer is used (Tonitto et al., 2006).

Cover crop benefits should be considered from a holistic point of view and not simply through a lens of one soil ecosystem service. For example, reduced water and wind erosion with cover crops can lead to reduced water pollution, reduced C losses, and increased soil fertility and productivity. The latter processes not only lead to sustained production of main crops but also support cover crop production for the continued delivery of soil ecosystem services. Such enhanced services from cover crops have ample positive agronomic, environmental, economic, and societal implications. Management of cover crops is key to their success for delivering soil ecosystem services.

16.1.2 Biomass Production

The amount of cover crop biomass production determines the extent to which cover crops can improve the services that soils provide. An increase in cover crop biomass production normally results in improved soil ecosystems services. Table 16.1 indicates cover crops can produce significant amounts of biomass under proper management, which has potential to exert significant changes in soil ecosystem services. Research also indicates the amount of cover crop biomass production can be variable due to various factors including cover crop management and climatic conditions.

While one cannot control climatic conditions, some cover crop management opportunities exist to reduce the variability in cover crop biomass production and further increase such production. Among the cover crop management opportunities include planting early and terminating late to extend the growing window, planting high-biomass producing species, increasing seeding rate, and irrigating and fertilizing cover crops (Blanco-Canqui, 2022; McClelland et al., 2021). Additional costs from increased seeding rate, irrigation, and fertilization could be offset by the greater availability of biomass for livestock or biofuel production as well as for the maintenance or improvement of other soil ecosystem services (Blanco-Canqui et al., 2020).

Planting a single species of high-biomass cover crops can be the most cost-effective opportunity to maintain or improve soil ecosystem services. Some may consider that cover crop mixes can produce more biomass, have greater resource use efficiency, and better adapt to changing environmental conditions than single species, but research data indicate mixes do not normally outperform single species cover crops (Florence & McGuire, 2020; MacLaren et al., 2019). Overall, biomass production could be manipulated with proper cover crop management under favorable weather conditions.

16.1.3 Economics

The benefits and goals of cover crops for improving soil ecosystem services are not really debatable. Who does not want to improve soil ecosystem services? However, economics often influence the decision on whether or not to adopt cover crops on a given farm and at large scales. Research data indicate grazing and harvesting cover crops can be one of the leading opportunities to generate positive net farm returns from cover crops if biomass production is sufficient to meet multiple goals from cover crops (Schomberg et al., 2014). The idea that grazing cover crops drastically compacts or damages soils and reduces crop yields is not supported by available research (Kelly et al., 2021). Grazing and harvesting cover crops does not generally erase the positive impacts of cover crops on ecosystem services.

Planting cover crops followed by grazing and harvesting is better than not planting cover crops to maintain or improve soil ecosystem services. This is because these practices do not remove the belowground biomass (roots) gained from planting cover crops in addition to a portion of aboveground cover crop biomass remaining after some biomass removal. The positive economic returns from grazing and harvesting cover crops could offset potential cover crop-induced reduction in subsequent crop yields, which preferentially occurs in water-limited regions (Holman et al., 2018).

Further, placing a monetary value on soil ecosystem services such as improved soil properties, reduced soil erosion, improved water quality, increased C sequestration, suppressed weeds, increased pollinator abundance, improved wildlife habitat (e.g., birds) is a missing opportunity to further boost net returns from cover crops (Dominati et al., 2010; Plastina et al., 2020; Roesch-McNally et al., 2018). Monetizing such essential services in addition to benefits to the whole society (e.g., clean water, reduced flood risks, landscape esthetics) from cover crops can incentivize cover crop adoption at larger scales to improve farm profitability while furthering the maintenance or improvement of soil ecosystem services.

16.1.4 Fluctuating Climates

An additional yet a crucial service from cover crops is their potential to help agricultural systems with adapting to current extreme weather events such as droughts, floods, heat waves, and intense rain and wind storms, among others (Kaye & Quemada, 2017). Experimental data indicate cover crops are effective at (a) reducing water erosion under high rainfall intensities and wind erosion under high wind speeds, (b) reducing evaporation and conserving water under moderate droughts, (c) improving drainage or infiltration in wet years, and (d) moderating abrupt fluctuations in soil temperature by reducing maximum and increasing minimum soil temperatures.

The ability of cover crops to contribute to extreme weather adaptation should not come as a surprise because cover crops were conceived to cover and protect the soil from changing weather conditions. Cover crops can equip the soil to resist abrupt changes in weather conditions and develop its resilience in the long term. Combining cover crops with no-till cropping systems can be a better option than combining with tilled systems as an adaption strategy to fluctuating climates. Indeed, cover crops can be particularly essential when transitioning from conventional till to no-till systems (Alletto et al., 2022). Cover crops can hasten the soil benefits of no-till after no-till adoption, facilitating adaptation to extreme weather events, considering that soil response to no-till without cover crops can be slow within the first few years after no-till adoption.

Depending on the severity of extreme events, insertion of cover crops can benefit from increased rainstorms. Indeed, research suggests cover crop biomass production can be increased under fluctuating climates due to increased intense precipitation, which in combination with the elevated CO_2, can enhance photosynthesis (Huang et al., 2020). The potential increase in biomass production under extreme conditions could synergistically boost the ability of agricultural lands to adapt and recover from stresses under further fluctuating climatic conditions.

16.2 Challenges

Published information indicates that while cover crops can provide numerous benefits and opportunities, their introduction does not come without significant challenges (Figure 16.2; Roesch-McNally et al., 2018; Plastina et al., 2020). Adding cover crops to a cropping system adds complexity to the system including possible changes in cropping systems, nutrient management, and additional field operations, labor, and costs (e.g., seed purchase). Also, the expected positive

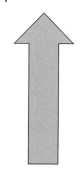

OPPORTUNITIES
1. Improves soil properties
2. Reduces soil erosion
3. Increases C stocks
4. Reduces nutrient leaching
5. Reduces the need for inorganic fertilizer
6. Suppresses weeds
7. Generates biomass for expanded uses

CHALLENGES
1. Increased production costs (e.g., seeds, labor)
3. Variable impacts on soil properties and soil C
4. Variable biomass production
5. Mixed effects on crop yields and economics

Figure 16.2 Cover crops can have more positive than negative impacts.

outcomes can be slow to develop or minimal in some cases. Assessing cover crop performance and benefits on a site-specific or local basis is critical to determine whether cover crops are delivering the intended benefits.

Positive impacts of cover crops observed in one soil or climatic region may not be observed in another environment with different soil, cover crop management, and climatic conditions. Cover crop performance in a given farm can even differ from the neighboring farm due to differences in cover crop management and soil-specific conditions. Potential benefits from cover crops are sometimes overemphasized in popular media, while challenges with cover crop implementation underemphasized. The emerging literature or research information provides an opportunity to discuss the potential challenges with cover crops.

16.2.1 Ecosystem Services

16.2.1.1 Biomass Production

Perhaps the single most important factor that determines cover crop effects on soil ecosystem service is the amount of cover crop biomass input as stressed throughout this book. Approximately 4 Mg ha^{-1} of cover crop biomass can be needed to significantly improve or enhance soil ecosystem services. Available data indicate that cover crops can produce significant amounts of biomass when averaged across all regions or studies, but the standard deviation of the reported means is large (Table 16.1; Ruis et al., 2019). Cover crop biomass production is highly variable from location to location and even within the same exact location from year to year. The variable biomass input can correspondingly result in variable impacts on soil ecosystem services.

The variability in biomass production is primarily due to differences in cover crop management and climate. Just as an example, in water-limited regions, success with rainfed cover crops depends on the year. In some years, cover crops do not germinate at all if precipitation is absent during the fallow period. Even in high precipitation regions, cover crop biomass production can be minimal if cover crops are planted late and terminated early to fit main crop harvest and planting times. For example, in the U.S. Midwest, winter cover crops planted in late fall after corn or soybean harvest and terminated in early spring of the following year often produce less than $1\,Mg\,ha^{-1}$ of biomass, which is too low to significantly modify soil ecosystem services (Koehler-Cole et al., 2020; Sindelar et al., 2019).

As discussed earlier, cropping systems that offer a long growing window for cover crops can be suitable for high biomass production. Such cropping systems include crop–fallow, corn silage, seed corn, short-season hybrids, specialty crops, and others harvested in summer or early fall (DeVincentis et al., 2020; Ruis et al., 2019). Delaying cover crop termination until main crops are planted can produce more biomass than when terminated early, but such strategy can result in reduced crop yields in years when precipitation input is low or zero at crop planting time.

16.2.1.2 Soil Properties

Most soil properties do not change rapidly after cover crop adoption. For instance, changes in soil physical properties can be slow to respond and not measurable in the short term even when biomass production is high. A review found that changes in some properties including bulk density, hydraulic conductivity, water retention (e.g., plant available water) parameters, and soil pore-size distribution did not occur, in some cases, even after 10 years (Blanco-Canqui & Ruis, 2020). Yet, changes in these and other soil properties determine the extent of services that soils provide. Also, incorporation of cover crop residues into the soil as green manure could diminish the positive effects of cover crops on soil properties, particularly soil structural properties (e.g., aggregate stability, number of biopores, water infiltration). Soil ecosystem services that primarily depend on surface coverage can rapidly improve after cover crop adoption. For example, well-established cover crops can immediately reduce soil erosion, buffer fluctuations in soil temperature, suppress weeds, and produce enough biomass for grazing or harvesting (Kelly et al., 2021). These rapid services from cover crops are reason enough to promote cover crop adoption, but improvement in most soil properties that affect long-term productivity may only manifest in the long term.

16.2.1.3 Carbon Sequestration

It is often touted in popular press that introducing cover crops into cropping systems can sequester large amounts of C in the soil. When experimental field data are considered, the potential of cover crops to alter soil organic C concentration and stocks is highly site-specific (Poeplau & Don, 2015; Blanco-Canqui, 2022). While global averages indicate cover crops can accumulate significant amounts of C (Table 16.1), cover crops may have small or no effects on soil C in some environments (Blanco-Canqui, 2022). The global averages do not imply that soils accumulate C at the reported averages in all soils or environments. The modest or small increments in soil C are commendable, but the idea that cover crops can lead to large amounts of C sequestered in all soils is not supported by the experimental data.

Amount of cover crop biomass C input, years after adoption, and initial C level affect the extent to which cover crops accumulate C. If biomass production is low and time after adoption is short, changes in soil C can be negligible or minimal (Blanco-Canqui, 2022). Also, while data on soil-profile C distribution under cover crops are limited, available data show that cover crops can primarily accumulate C in the shallow surface (<20 cm depth). The shallow C can be susceptible to losses via emissions and erosion. The challenge is to implement cover crops that can transfer C to deeper depths and promote deep C stability or sequestration.

Interest in growing cover crops is particularly high in cool temperate environments with highly fertile and high initial C soils. However, these soils may not benefit from cover crop as much as soils in warm environments or degraded soils with low fertility and low initial C. Degraded soils could accumulate larger amounts of C due to their lower C saturation when cover crops are successfully introduced and managed. The challenge is that degraded soils are often degraded for a reason. Restoring degraded soils may not be an easy task. These soils are often found in water-limited environments or marginally-productive agricultural lands with low cover crop biomass production potential. These challenges underscore the need for site-specific assessment and development of cover crop practices to enhance C sequestration.

16.2.1.4 Dissolved Nutrients

One of the traditional services from cover crops is the reduction in water erosion. Often this positive service is equated to significant reduction in runoff, sediment loss, sediment-associated nutrients, and dissolved nutrients in runoff. Research data show that cover crops can reduce sediment loss and sediment-associated nutrients, but their potential to reduce losses of dissolved nutrients in runoff can be limited. Runoff leaving fields with cover crops can have lower sediment concentration than that leaving fields without cover crops but can still carry significant amount of dissolved nutrients (Liu et al., 2019). The limited effectiveness of cover crops to reduce concentrations of dissolved nutrients in runoff can be a challenge when cover crops are promoted as an effective practice to reduce non-point source pollution and improve water quality. Also, in cool temperate regions, cover crop residues can release nutrients such as P due to intense freeze–thaw cycles (Liu et al., 2019). Thus, concentration of dissolved P in runoff from fields with cover crops can be greater than from fields without cover crops although the cover crop-derived dissolved P losses in runoff can be relatively small when compared to the reduction of total P after adding cover crops. A potential strategy to augment cover crop potential to reduce dissolved nutrient losses can be the use of companion practices such as conservation buffers and adherence to the 4Rs guidelines.

16.2.1.5 Crop Yields

How cover crops affect crop yields is debatable (Deines et al., 2023). Cover crops can have positive, negative, and no effects on crop yields. The mixed effects of cover crops on yields are a challenge for the widespread adoption of cover crops. The potential positive effects of cover crops on soil ecosystem services (e.g., reduced soil erosion) not directly related to crop yields may not be sufficient to convince producers to adopt cover crops unless costs associated with the potential decrease in yields are offset. Cover crops such as non-legumes can either reduce or have no effects on crop yields. In humid regions, non-legume cover crops can reduce crop yields by immobilizing nutrients, reducing seed emergence and establishment of the main crops (Marcillo & Miguez, 2017; Deines et al., 2023).

In water-limited environments, cover crops can reduce subsequent crop yields by using the limited amount of water. Data show cover crops can reduce subsequent crop yields in approximately 50% of cases in water-limited regions (Blanco-Canqui et al., 2022). Yield reduction with cover crops can be particularly large in dry years or when precipitation amount is below normal. In water-limited regions, the revolving challenge is how to grow cover crops that produce significant amounts of biomass for contributing to soil ecosystem services without reducing crop yields.

16.2.1.6 Economics

Generating immediate positive net returns from cover crops is still uncertain although subsidies are available in some regions to incentivize cover crop adoption (Bergtold et al., 2017; Plastina et al., 2020; Prabhakara et al., 2015). While grazing and harvesting of cover crops are highly promising to improve farm economics, this opportunity can come with challenges. Limited knowledge and time for cover crop and livestock management could complicate this opportunity. Selecting the right cover crop species and maximizing biomass production can be key to simultaneously support both soil conservation and livestock production goals (Kelly et al., 2021). Also, challenges exist with monetization of soil ecosystem services (e.g., C credits) to generate profits and entice large-scale adoption of cover crops. Cover crop establishment (e.g., seeds, planting, termination) can be costly and rapid return on investment rightly expected. Cover crops are a long-term investment as mentioned earlier. Some of the benefits from cover crops are external to farmers such as improved water and air quality, which may not provide immediate net returns unless commodified based on the benefits to the whole society.

16.2.1.7 Fluctuating Climates

Extreme weather-proofing of agricultural lands with cover crops can be lofty goal but may not be easily achievable in all scenarios. The effectiveness of cover crops for adaptation to extreme weather events depends on the amount of biomass produced. The challenge is that there is a mutual relationship between cover crop biomass production and extreme weather events. Frequent droughts or floods can prevent cover crop emergence and biomass production. Well-established cover crops are needed to buffer the extreme events.

Long-term use of cover crops can be critical to enhance soil resilience against extreme events. For example, improving soil structural properties to improve water movement and storage and increasing soil organic matter concentration for improving related soil properties can take years before significant changes are measurable. Cover crops might be able to help with adaptation to moderate droughts, wet conditions, and erosion, but their effectiveness for adaptation to severe events can be limited. Indeed, in some cases, cover crops could aggravate effects of extreme events such as droughts by depleting soil water needed for the next crops.

16.3 Remaining Questions

The abundant literature on cover crops has advanced our understanding of the opportunities and challenges with cover crop management. Some of the remaining gaps that need to be addressed to further advance our knowledge of cover crops include:

1) Additional data from long-term (>10 years) experiments on cover crops and all soil ecosystem services are needed. Despite decades if not centuries of cover crop research, most data are from 3- to 5-year studies. Also, most studies focus on one or two ecosystem services and not all the relevant services. Coordinated research across disciplines comprising soil scientists, climatologists, animal scientists, plant scientists, economists, hydrologists, ecologists, and others can help to more comprehensively examine broad implications of cover crop adoption on soil ecosystem services that affect the whole society.
2) Our current knowledge of the potential of cover crops to sequester C in the soil profile is incomplete as most available data on soil C are from the shallow surface (<30 cm depth). Data on soil-profile C distribution as affected by cover crops are needed for an accurate inventory of C stocks. The hypothesis is, in high

precipitation regions, soil C could be translocated to deeper depths in the long term, contributing to deep C sequestration. Also, in temperate soils, soil C is often stratified near the surface; but in tropical regions, significant translocation of C to the subsoil is common. For example, in a degraded tropical soil in Brazil, only half of soil C under cover crops was stored in the 0 to 30 cm depth, while the rest was stored in the 30–100 cm depth (Veloso et al., 2018). In a Mediterranean climate in the U.S., cover crops accumulated more C in the 0–30 cm depth but stored less C in the 30–200 cm depth relative to no cover crops, resulting in overall C losses for the whole profile (Tautges et al., 2019). These examples highlight the need for a soil-profile C inventory for a complete understanding.

3) Commodification of essential ecosystems services improved by cover crops can be an option to improve farm economics and incentivize adoption of cover crops. Placing a monetary value on soil ecosystem services (e.g., C sequestered, improved water quality) that affect the whole society could increase total net returns in addition to direct returns from grazing and harvesting cover crops. Commodification of C in trading markets is developing, but more long-term soil C data are still needed for this purpose. A comprehensive assessment of direct and indirect (e.g., hidden) costs and benefits from cover crops is unavailable. Yet, this is needed if adoption of cover crops at larger scales is a goal.

4) Significant amounts of cover crop biomass is needed to improve soil properties, reduce erosion, suppress weeds, sequester C, and provide other benefits, but biomass production is often limited or highly variable. Additional research on strategies to maximize cover crop biomass production without adversely affecting other services such as crop production is much needed. Optimum cover crop planting and termination dates, seeding rates, and planting methods (e.g., aerial seeding, drilling) are yet to be refined for different climatic regions to achieve multiple and enhanced services from cover crops.

5) Studies reporting cover crop biomass production abound but those reporting belowground cover crop biomass production (roots) are few. Yet, root biomass production can be equally, if not more important than, aboveground biomass production for improving soil ecosystem services. Characterization of soil-profile root biomass production under different cover crop species, management settings, soil type, and climates is needed to better discern root-soil interactions for the whole soil profile associated with soil ecosystem services.

6) Adapting against fluctuating weather patterns by adopting cover crops sounds promising. However, specific research information on the extent to which cover crops can contribute to the adaptation of agroecosystems to different intensities of droughts, rainstorms, windstorms, extreme soil temperatures, floods, and other climatic events is still limited. Simulation studies under different levels of weather intensities (e.g., low, moderate, and severe droughts) can provide useful information.

7) The emerging cover crop breeding efforts should be further amplified to develop species that are tolerant and resistant to variable environmental conditions (e.g., salinity, low fertility) and extreme weather events (e.g., droughts, heat waves). Because cover crop biomass production determines cover crop potential for altering soil ecosystem services, cover crops that produce generous amounts of aboveground and belowground (root) biomass in a short period of time or short growing season can be ideal to maintain or boost soil ecosystem services.

8) Grazing and harvesting cover crops are emerging as top strategies to offset cover crop establishment costs and improve farm profits. However, most of the available data on cover crop grazing and harvesting are from short-term (<5 years) experiments. How long-term grazing and harvesting of cover crops affect soil ecosystems services is poorly understood. More comprehensive assessment of cover crop biomass removal under different animal stocking rates, grazing duration, amount of biomass removal, climatic conditions, and soil conditions (e.g., wet, dry, frozen) is needed.

9) Cover crops have generated much interest and excitement. However, many have established cover crops on croplands with ideal soils without much consideration of the initial soil fertility or C level. Indeed, most research data on cover crops are from cool temperate regions with high initial soil C levels or high fertility soils. Data on cover crop performance from degraded environments with low initial soil C levels, low fertility, sandy, eroded, and marginal soils are few. The hypothesis is that cover crops could improve soil ecosystems services more in problem soils than in soils with few agronomic and environmental limitations.

16.4 The Future of Cover Crops

The future for cover crops is highly promising. Opportunities discussed above could outweigh the challenges. Cover crop technology has a unique potential to address current concerns on declining soil ecosystem services and increasing extreme weather events (Fig. 16.1). The potential improvement in soil properties may not only restore soil ecosystem services but also improve soil resilience against fluctuating weather conditions. Cover crops can be a critical piece within the portfolio of options to develop extreme weather-resilient systems that can minimize impacts of droughts, floods, heat waves, rain and wind storms, and snowstorms, among others.

However, this promising future of cover crops to address concerns of broad societal implications will depend on the extent of cover crop adoption and how cover crops are managed. Cover crop performance is highly site-specific as repeatedly mentioned in each chapter of this book. It is keenly important to recognize that the multi-functionality of cover crops is dependent on cover crop biomass production. Site-specific strategies will be needed to refine cover crop management to ensure adequate biomass production for restoring the declining soil ecosystem services.

Success with cover crops not only depends on land availability or cover crop management skills but also on economic and policy issues. Adoption may continue to be slow if no economic benefit is attained from planting cover crops. Experimental data showing no adverse impacts of grazing and harvesting on soils and crop yields are a favorable outlook to attain net returns from cover crops. Additionally, some payment for soil ecosystem services improved by cover crops can provide economic returns while promoting large-scale adoption of cover crops. Such payment can be especially fair when cover crops have potential to deliver services of societal value such as improved environmental quality and adaptation to current extreme weather events. Also, restoration of soil C lost with cover crops is an emerging opportunity for economic gains via C trading. Clear guidelines for cover crop management along with adoption options (e.g., cost-share, insurance) can help with cover crop integration into current cropping systems.

Cover crops were initially designed to reduce soil erosion, help with soil fertility, and suppress weeds. Now, available research information support the notion cover crops could be able to deliver even more services when managed properly through refinement of local cover crop management practices. As per the list under the remaining questions above, innovating and refining management strategies such as developing optimum planting and termination dates, integrating with livestock to generate economic returns, monetizing some ecosystem services of societal value await attention.

Cover crops may not address all soil problems, but they can be an essential companion practice to no-till, crop residue management, conservation buffers and other conservation practices to improve or enhance soil ecosystem services. While challenges exist, cover crops have more positive than negative effects. Thus, planting cover crops is better than not planting cover crops. Cover crops should be considered as an integral component of current management scenarios to address the increasing challenges with declining soil resilience and ecosystem services under variable weather patterns considering the opportunities can outweigh the challenges.

16.5 Summary

A number of opportunities and challenges exist with cover crop adoption. Cover crops can provide multiple ecosystem services when biomass production is sufficiently high. Such benefits include improved soil properties, reduced soil erosion, improved nutrient retention, and increased C sequestration, soil fertility, weed suppression, and biomass production. An ancillary positive outcome from cover crops can be their contribution to adaptation to fluctuating climates. Also, cover crops provide opportunities to increase farm profits via grazing and harvesting in addition to potential monetization of soil ecosystem services (e.g., C credits, credits for improved water quality).

One of the major challenges with cover crops is the variable biomass production, which can directly determine their soil ecosystem service benefits. Another challenge is the potential reduction in crop yields due to water use by cover crops, particularly in water-limited environments. To address such challenges, upcoming research should: (a) develop management strategies that maximize cover crop biomass production to achieve multiple soil services and minimize impacts on subsequent crop yields, (b) gather long-term (>10 years) data reporting all soil ecosystem services via multi-disciplinary research, (c) characterize soil C for the whole soil profile for an accurate inventory needed for C trading programs, (d) evaluate the potential monetization of all essential soil ecosystem services, and (e) focus on environments where cover crops can provide greater benefits such as degraded soils with low fertility and low initial C levels. In conclusion, research data indicate cover crops generally improve soil ecosystem services, help with extreme weather adaptation, and potentially increase farm economics.

References

Abdalla, M., Hastings, A., Cheng, K., Yue, Q., Chadwick, D., Espenberg, M., Truu, J., Rees, R. M., & Smith, P. (2019). A critical review of the impacts of cover crops on nitrogen leaching, net greenhouse gas balance and crop productivity. *Global Change Biology, 25*, 2530–2543.

Adetunji, A. T., Ncube, B., Mulidzi, R., & Lewu, F. B. (2020). Management impact and benefit of cover crops on soil quality: A review. *Soil and Tillage Research, 204*, 104717.

Alletto, L., Cassigneul, A., Duchalais, S., Giuliano, J., Brechemier, J., & Justes, E. (2022). Cover crops maintain or improve agronomic performances of maize monoculture during the transition period from conventional to no-tillage. *Field Crops Research, 283*, 108540.

Bergtold, J. S., Ramsey, S., Maddy, L., & Williams, J. R. (2017). A review of economic considerations for cover crops as a conservation practice. *Renewable Agriculture and Food Systems, 34*, 62–76.

Blanco-Canqui, H. (2022). Cover crops and carbon sequestration: Lessons from US studies. *Soil Science Society of America Journal, 86*, 501–519.

Blanco-Canqui, H., & Ruis, S. (2020). Cover crops and soil physical properties. *Soil Science Society of America Journal*, 1527–1576.

Blanco-Canqui, H., Ruis, S., Holman, H., Creech, C., & Obour, A. (2022). Can cover crops improve soil ecosystem services in water-limited environments? A review. *Soil Science Society of America Journal, 86*, 1–18.

Blanco-Canqui, H., Ruis, S., Proctor, C., Creech, C., Drewnoski, M., & Redfearn, D. (2020). Harvesting cover crops for biofuel and livestock production: Another ecosystem service? *Agronomy Journal, 112*, 2373–2400.

Deines, J. M., Guan, K., Lopez, B., Zhou, Q., White, C. S., Wang, S., Lobell, D. B. (2023). Recent cover crop adoption is associated with small maize and soybean yield losses in the United States. *Global Change Biology, 29*, 794–807.

DeVincentis, A. J., Solis, S. S., Bruno, E. M., Leavitt, A., Gomes, A., Rice, S., & Zaccaria, D. (2020). Using cost-benefit analysis to understand adoption of winter cover cropping in California's specialty crop systems. *Journal of Environmental Management, 261*, 110205.

Dominati, E., Patterson, M., & MacKay, A. (2010). A framework for classifying and qualifying the natural capital and ecosystem services of soils. *Ecological Economics, 69*, 1858–1868.

Florence, A. M., & McGuire, A. M. (2020). Do diverse cover crop mixtures perform better than monocultures? A systematic review. *Agronomy Journal, 112*, 3513–3534.

Holman, J. D., Arnet, K., Dille, J., Maxwell, S., Obour, A., Roberts, T., Roozeboom, K., & Schlegel, A. (2018). Can cover or forage crops replace fallow in the semiarid central Great Plains? *Crop Science, 58*, 932–944.

Huang, Y., Ren, W., Grove, J., Poffenbarger, H., Jacobsen, K., Tao, B., Zhua, X., & McNear, D. (2020). Assessing synergistic effects of no-tillage and cover crops on soil carbon dynamics in a long-term maize cropping system under climate change. *Agricultural and Forest Meteorology, 291*, 108090.

Kaye, J., & Quemada, M. (2017). Using cover crops to mitigate and adapt to climate change. A review. *Agronomy for Sustainable Development, 37*, 4.

Kelly, C., Schipanski, M. E., Tucker, A., Trujillo, W., Holman, J. D., Obour, A. K., Johnson, S. K., Brummer, J. E., Haag, L., & Fonte, S. J. (2021). Dryland cover crop soil health benefits are maintained with grazing in the U.S. High and Central Plains. *Agriculture, Ecosystems and Environment, 313*, 107358.

Koehler-Cole, K., Elmore, R. W., Blanco-Canqui, H., Francis, C. A., Shapiro, C. A., Proctor, C. A., Heeren, D. M., Ruis, S., Irmak, S., & Ferguson, R. B. (2020). Cover crop productivity and subsequent soybean yield in the western Corn Belt. *Agronomy Journal, 112*, 2649–2663.

Liu, J., Macrae, M. L., Elliott, J. A., Baulch, H. M., Wilson, H. F., & Kleinman, P. J. A. (2019). Impacts of cover crops and crop residues on phosphorus losses in cold climates: A review. *Journal of Environmental Quality, 48*, 850–868.

MacLaren, C., Swanepoel, P., Bennett, J., Wright, J., & Dehnen-Schmutz, K. (2019). Cover crop biomass production is more important than diversity for weed suppression. *Crop Science, 59*, 733–748.

Marcillo, G. S., & Miguez, F. E. (2017). Corn yield response to winter cover crops: An updated meta-analysis. *Journal of Soil and Water Conservation, 72*, 226–239.

McClelland, S. C., Paustian, K., & Schipanksi, M. E. (2021). Management of cover crops in temperate climates influences soil organic carbon stocks - A meta-analysis. *Ecological Applications, 31*, e02278.

Osipitan, O. A., Dille, J. A., Assefa, Y., & Knezevic, S. Z. (2018). Cover crop for early season weed suppression in crops: Systematic review and meta-analysis. *Agronomy Journal, 110*, 2211–2221.

Plastina, A., Liu, F., Miguez, F., & Carlson, S. (2020). Cover crops use in Midwestern US agriculture: Perceived benefits and net returns. *Renewable Agriculture and Food Systems, 35*, 38–48.

Poeplau, C., & Don, A. (2015). Carbon sequestration in agricultural soils via cultivation of cover crops – A meta-analysis. *Agriculture, Ecosystems and Environment, 220*, 33–41.

Prabhakara, K., Hively, W. D., & McCarty, G. W. (2015). Evaluating the relationship between biomass, percent groundcover and remote sensing indices across six winter cover crop fields in Maryland, United States. *International Journal of Applied Earth Observation and Geoinformation, 39*, 88–102.

Roesch-McNally, G., Basche, A., Arbuckle, J., Tyndall, J., Miguez, F., Bowman, T., & Clay, R. (2018). The trouble with cover crops: Farmers' experiences with overcoming barriers to adoption. *Renewable Agriculture and Food Systems, 33*, 322–333.

Ruis, S. J., Blanco-Canqui, H., Creech, C. F., Koehler-Cole, K., Elmore, R. W., & Francis, C. A. (2019). Cover crop biomass production in temperate agroecozones. *Agronomy Journal, 111*, 1535–1551.

Schomberg, H., Fisher, D., Reeves, D., Endale, D., Raper, R., Jayaratne, K., Gamble, G., & Jenkins, M. (2014). Grazing winter rye cover crop in a cotton no-till system: Yield and economics. *Agronomy Journal, 106*, 1041–1050.

Sindelar, M., Blanco-Canqui, H., Virginia, J., & Ferguson, R. (2019). Cover crops and corn residue removal: Impacts on soil hydraulic properties and their relationships with carbon. *Soil Science Society of America Journal, 83*, 221–231.

Tautges, N. E., Chiartas, J. L., Gaudin, A. C. M., O'geen, A. T., Herrera, I., & Scow, K. M. (2019). Deep soil inventories reveal that impacts of cover crops and compost on soil carbon sequestration differ in surface and subsurface soils. *Global Change Biology, 25*, 3753–3766.

Tonitto, C., David, M. B., & Drinkwater, L. E. (2006). Replacing bare fallows with cover crops in fertilizer-intensive cropping systems: A meta-analysis of crop yield and N dynamics. *Agriculture, Ecosystems and Environment, 112*, 58–72.

Veloso, M. G., Angers, D. A., Tiecher, T., Giacomini, S., Dieckow, J., & Bayer, C. (2018). High carbon storage in a previously degraded subtropical soil under no tillage with legume cover crops. *Agriculture, Ecosystems and Environment, 268*, 15–23.

Appendix I

Common and Scientific Names Used in the Book

Alfalfa	*Medicago sativa*
Austrian pea	*Pisum sativum* L.
Barley	*Hordeum vulgare* L.
Bean	*Phaseolus vulgaris* L.
Berseem clover	*Trifolium alexandrinum*
Black oat	*Avena strigosa*
Blue lupine	*Lupinus angustifolius*
Bristle oat	*Avena strigosa*
Brown mustard	*Brassica juncea*
Buckwheat	*Fagopyrum esculentum*
Cahaba vetch	*Vicia spp.*
Camelina	*Camelina sativa*
Canola	*Brassica napus*
Carpet grass	*Axonopus compressus*
Clover	*Trifolium*
Corn	*Zea mays*
Creeping grass	*Cynodon plectostachyum*
Crimson clover	*Trifolium incarnatum*
Dwarf pigeon pea	*Cajanus cajan*
Elephant grass	*Pennisetum polystachion*
Field peas	*Pisum sativum*
Field pennycress	*Thlaspi arvense*
Grain sorghum	*Sorghum bicolor*
Guinea grass	*Panicum maximum*
Hairy vetch	*Vicia villosa*
Horseweed	*Erigeron canadensis*

Cover Crops and Soil Ecosystem Services, First Edition. Humberto Blanco.
© 2023 American Society of Agronomy, Inc. / Crop Science Society of America, Inc.
/ Soil Science Society of America, Inc. Published 2023 by John Wiley & Sons, Inc.

Jack bean	*Canavalia ensiformis*
Kochia	*Bassia scoparia*
Kudzu	*Pueraria phaseoloides*
Lupine	*Lupinus*
Millet	*Pennisetum glaucum*
Mustard	*Brassica*
Oat	*Avena sativa*
Oil radish	*Raphanus sativus*
Phacelia	*Phacelia tanacetifolia*
Potato	*Solanum tuberosum*
Purple top turnip	*Brassica campestris*
Radish	*Raphanus sativus*
Rapeseed	*Brassica napus*
Red clover	*Trifolium pratense*
Rice	*Oryza sativa*
Rye	*Secale cereale*
Ryegrass	*Lolium*
Showy rattlebox	*Crotalaria spectabilis* Fabales
Soybean	*Glycine max*
Style	*Stylosanthes gracilis*
Sudan grass	*Sorghum* × *drummondii*
Sunn hemp	*Crotalaria juncea*
Sweet clover	*Melilotus officinalis*
Timothy	*Phleum pratense*
Triticale	×*Triticosecale* Wittmack
Turnip rape	*Brassica rapa*
Velvet bean	*Mucuna pruriens*
Wheat	*Triticum aestivum*

Printed and bound by CPI Group (UK) Ltd, Croydon, CR0 4YY
20/08/2023

08102009-0001